# Advancing Global Bioethics

Volume 21

The book series Global Bioethics provides a forum for normative analysis of a vast range of important new issues in bioethics from a truly global perspective and with a cross-cultural approach. The issues covered by the series include among other things sponsorship of research and education, scientific misconduct and research integrity, exploitation of research participants in resource-poor settings, brain drain and migration of healthcare workers, organ trafficking and transplant tourism, indigenous medicine, biodiversity, commodification of human tissue, benefit sharing, bio-industry and food, malnutrition and hunger, human rights, and climate change.

Lillian Omutoko • Walter Jaoko
Editors

# Bioethics from the Global South

Springer

*Editors*
Lillian Omutoko
Department of Educational Management,
Policy and Curriculum Studies
University of Nairobi
Nairobi, Kenya

Walter Jaoko
KAVI-Institute of Clinical Research
University of Nairobi
Nairobi, Kenya

ISSN 2212-652X ISSN 2212-6538 (electronic)
Advancing Global Bioethics
ISBN 978-3-031-77668-7 ISBN 978-3-031-77669-4 (eBook)
https://doi.org/10.1007/978-3-031-77669-4



This Springer imprint is published by the registered company Springer Nature Switzerland AG
The registered company address is: Gewerbestrasse 11, 6330 Cham, Switzerland

*Dedicated to all those who have been instrumental in shaping our journey in research ethics and all those working in the research ethics and bioethics space.*

# Foreword

As the academic coordinator of the Master in Global Bioethics, I have had the privilege to know most of the authors. It is with great pleasure that I write these words. Global Bioethics has developed rapidly, as Dr. Henk Ten Have explains in his book: "The global dimension of bioethical issues has produced a new stage, or a new kind of bioethics. First, Global Bioethics has become a worldwide discipline since it attends to the ethical concerns of humanity, no longer limited to a single country or region of the world, or a particular part of the world population. Second, global bioethics is encompassing, as it considers the ethical values and principles of various populations without assuming that one specific set of values and principles is dominant" (Have, Henk Ten; Global Bioethics: An Introduction 2016).

The discourse of bioethics has long been dominated by voices from the global north. Ethical debates, research, and policy discussions often revolve around issues and challenges primarily relevant to affluent nations, leaving the unique perspectives and concerns of the developing world in the shadows. It is within this context that the book "Bioethics from the Global South" emerges as a groundbreaking and much-needed contribution to the field with a different perspective. This remarkable work, a collaboration of 11 distinguished bioethicists, showcases the remarkable dedication and academic skillfulness of individuals who share a common academic formation. Nine of the authors are alumni of the Master in Global Bioethics program at the School of Bioethics, Anahuac University, a testament to the program's commitment to the mission of the University: "the comprehensive training of positive action leaders and institutionally promoting the development of the person and society, inspired by the values of Christian humanism."

"Bioethics from the Global South" is not just another book on bioethics; it is a pioneering effort that challenges the traditional boundaries of ethical discourse. In a world where globalization has intertwined our destinies and interconnected our challenges, the need for a global bioethical framework becomes evident. This book explores a myriad of ethical issues that transcend borders, including Climate Change, Environmental Ethics, Reproductive Health and Technologies, Surrogacy, Global Justice, Ethics of Vaccine Distribution, Public Health Ethics, and COVID-19.

The COVID-19 pandemic has starkly illustrated that bioethical issues know no national or regional boundaries. It has brought to the forefront the imperative need for a global perspective in addressing ethical challenges in healthcare, research, and public health. The authors seize this moment to remind us that ethical questions must be tackled collectively and comprehensively, transcending the confines of any single nation or culture, and taking into consideration many factors and different situations. It emphasizes that one policy cannot be enforced without taking into consideration the human rights of the person.

As I read into the chapters, I discovered a myriad of bioethical issues that often remain overlooked. This book gave voice to those often unheard, shedding light on matters rarely discussed within the broader context of global bioethics. It was a refreshing and enlightening experience. The personalist approach to bioethics, which I teach, underscores the sacredness of every life and the importance of preserving human dignity. This perspective resonates throughout the book, guiding the ethical discussions and framing the authors' analyses of complex global bioethical dilemmas. "Bioethics from the Global South" is not only an academic endeavor but also a call to action. It highlights the challenges faced by global bioethics, such as cultural diversity, resource limitations, new technologies, and the need for greater awareness and education in low- and middle-income countries. It emphasizes the importance of universal frameworks, like the Universal Declaration of Human Rights and the Universal Declaration of Bioethics and Human Rights, in guiding ethical discussions and decision-making on a global scale.

This book is a testament to the power of collaboration and the strength that comes from diverse voices coming together to address shared challenges. The authors, hailing from countries such as Kenya, Uganda, and Nigeria, bring a rich tapestry of experiences and perspectives to the global bioethics discourse, reminding us that every culture has value and that every traditional value needs to be brought to the table so that, together, policies and decisions can be made that address the needs of the community as well as the country. Citing our university motto, "Vince Bono Malum," which means "Overcome Evil with Good," this book represents a positive action program to overcome the evil of ignorance with the good of science and knowledge. The book is a significant contribution to the field of bioethics, one that will undoubtedly stimulate further dialogue and research. I extend my heartfelt congratulations to the authors for their outstanding work in bringing the global south's bioethical perspectives to the forefront. May this book serve as an inspiration for continued dialogue and action in the pursuit of a more ethical and equitable world for all.

Atlanta, Georgia, USA
2024

Maria Ines Girault

# Preface

This is an intersection of ethics, humanity, and diverse bioethical concerns of the global south. The intellectual journey recognizes that our paths point to academic inquiry and the well-being of individuals, communities, ecosystems, and future generations. The ethical dilemmas faced by the global south are unique, shaped by population increase, climate change, a plethora of public health emergencies, globalization, and technological advancement. This book delves into the ethical intricacies of public health ethics, health emergencies, environmental ethics, and reproductive technologies. Moral discourse has been engaged in the fields of health, health-related challenges, epidemics, pandemics, and emergencies. It is evident that bioethical issues have no national or regional boundaries, and thus there is a need for global collective responsibility in dealing with public health, emergencies, the environment, surrogacy, and medical migration.

Scientific and technological advances have changed the way vulnerabilities of nature are handled bringing forth a myriad of changes that affect human lives. Global bioethics serves to save, improve, and extend human lives. The moral concerns raised therein have been tackled from a bioethical perspective with chapters drawn from Asia and Africa. The discourse begins with acknowledging ethical dilemmas, then involves ethical analyses before examining the complex global problems through bioethics theoretical models. The principlism approach, a universal principle-based ethics, has been applied to cases irrespective of geographical location and context and used to examine moral dilemmas and decision-making. A global bioethics framework has been used to analyze the ethical issues that shake and test the moral fabric of society when healthcare and well-being are threatened; when disaster strikes; when the planet is disturbed; and when bodies become vessels and dreams take shape.

This Framework provides a lens for evaluating and solving global bioethical problems. The Universal Declaration of Human Rights, Universal Declaration of Bioethics and Human Rights; and respective national constitutions were used to argue for the right to human dignity, health, and other fundamental freedoms. Comprehensiveness of the framework provides space to incorporate traditions, values, and cultures, and provides a basis for examining ethical challenges such as vulnerability, equity, justice, solidarity, and cooperation. The precautionary principle was proposed as suitable for states to take social responsibility to protect the public from exposure to harm occasioned by activities that have been found to have plausible risks. This principle recommends that policy or decision-makers take precautions to protect the environment and human health.

Bioethics from the Global South is a culmination of documentary research done by postgraduate students of Global Bioethics at Anahuac University and academic scholars with a keen interest in Bioethics. The book has a wide perspective, and the authors are from various countries including Kenya, Uganda, and Nigeria. The power of like minds is testimony to the possibility of challenging assumptions, embracing complexities, and seeking solutions that honor human dignity and ecological balance.

Nairobi, Kenya

Lillian Omutoko
Walter Jaoko

# Acknowledgments

We extend our deepest gratitude to everyone who supported the writing of this book. Special gratitude to Anahuac University, Mexico, for being instrumental in shaping the author's thoughts on bioethics discourse in the global south. We are grateful to the affiliate institutions for providing the authors with scholarly environments and opportunities to engage in academic endeavors. We appreciate the esteemed authors for the commitment to make the book a success.

This book is an offshoot of the Masters in Global Bioethics, thank you Dr. Maria Girault for walking with most of the authors in the journey of Global Bioethics. The program opened our eyes and enables us to see and hear what we otherwise would not see or hear. Truly, we are indebted to all the Faculty of Anahuac University who walked us through very intriguing topical areas, some of which were amazing and even surprising. Prof. Henk Ten Have, thank you for urging us on and giving much-needed advice, especially during the conceptualization of the idea, without which this book may never have seen the light of day. Prof. Karori Mbugua, we appreciate the role you played in evaluating the chapters.

To the reviewers and external editors, thank you, we could not have done it without you.

Lillian Omutoko
Walter Jaoko

# Contents

# Contributors

**Walter Jaoko** KAVI-Institute of Clinical Research, University of Nairobi, Nairobi, Kenya

**Teresia Nyawira Kaniaru** National Commission for Science Technology and Innovation, Nairobi, Kenya

**Melba Katindi** Katindi & Co. Advocates, Nairobi, Kenya

**Everlyn Kisembe** School of Arts and Social Sciences, Moi University, Kesses, Kenya

**Daniel Lagat** Department of Education, School of Business and Education, Kirinyaga University, Kerugoya, Kenya

**Eva Miria Magambo** Uganda National Association of Community and Occupational Health (UNACOH), Kampala, Uganda

**Karori Mbũgua** Department of Philosophy and Religious Studies, University of Nairobi, Nairobi, Kenya

**Violet Naanyu** School of Arts and Social Sciences, Moi University, Kesses, Kenya

**Anthony Kebira Nyamache** Department of Biochemistry, Microbiology and Biotechnology, Kenyatta University, Nairobi, Kenya

**Emmanuel E. Okenwa-Vincent** Department of Optometry and Vision Sciences, Kaimosi Friends University, Kaimosi, Kenya

**Lillian Omutoko** Department of Educational Management, Policy and Curriculum Studies, University of Nairobi, Nairobi, Kenya

**Mercury Shitindo** Africa Bioethics Network, Nairobi, Kenya

Health Sciences, University of Zaragoza, Zaragoza, Spain

**Jane Wathuta** Strathmore Law School, Strathmore University, Nairobi, Kenya

# Part I
# Public Health Ethics, COVID-19, Ethics of Vaccine Distribution, Uptake and Governance

# Chapter 1
# Adherence to the Standard Operating Procedures to Prevent Spread of COVID-19 in Nakawa Division, KCCA, Uganda: Reality or Myth? An Exploratory Study

**Eva Miria Magambo**


## Abstract

**Introduction**
COVID-19 continues to ravage society. Globally, 178,503,429 COVID-19 cases were confirmed including 3,872,457 deaths by June 22, 2021, WHO 2021. In Uganda, from January 3, 2020, to June 22, 2021, 72,679 cases of COVID-19 were confirmed with 680 deaths. COVID-19 is caused by SARS-CoV-2 (CDC, Basics of COVID-19; Updated May 24, 2021. CDC, Atlanta, 2021). While disease is preventable, change in human behavior, which ranks high among preventive measures, presents ethical–moral problems. Uganda experienced a second wave of the COVID-19 pandemic with a sevenfold increase in the number of COVID-19 cases and an escalation in deaths.

**Objectives**
The study explored adherence to SOPs and was motivated by the problem of nonadherence to COVID-19 prevention measures: proper wearing of masks, handwashing with soap, water/use of hand sanitizers, and keeping social distance.

**Methods**
Documentary research project information resources were sourced online, using research methodology tools. Data was analyzed, synthesized, and compared using an interdisciplinary approach, and a report was compiled.


E. M. Magambo (✉)
Uganda National Association of Community and Occupational Health (UNACOH), Kampala, Uganda



L. Omutoko, W. Jaoko (eds.), *Bioethics from the Global South*, Advancing Global Bioethics 21, https://doi.org/10.1007/978-3-031-77669-4_1

**Results**
Several dimensions of nonadherence to SOPs were elaborated, which included a historical perspective, law, scientific aspects, medicine, technology, socioeconomics, politics/political science, policy, anthropology, sociology, psychology, culture, ethnicity, theology, demographics, human behavior, environment, information, education communication, community development, and mobilization.

**Conclusions/Recommendations**
Nonadherence to SOPs was widely demonstrated and noted as a major reason for the worsening pandemic in Uganda. Study results recommend an interdisciplinary response to mitigate the spread of pandemics.

**Keywords** Adherence · Behavior · COVID-19 · Multidimensional · Pandemic · Standard operating procedures

## 1 Introduction and Background

The World Health Organization (WHO) declared COVID-19 a pandemic on March 11, 2020 (Cucinotta & Vanelli, 2020). It was first reported in China in December 2019 (Rodriguez-Morales et al., 2020); since then the disease continued to ravage the entire globe. Twenty million cases of COVID-19 and 738,937 deaths were reported globally by August 2020 (Manikandan, 2020), while WHO reported 178,503,429 COVID-19 cases and 3,872,457 deaths as of June 2021. Vaccine doses totaling 2,414,347,324 were administered globally by June 22, 2021. WHO further reported that in Uganda, from January 3, 2020, to June 22, 2021, 72,679 COVID-19 cases had been confirmed with 680 deaths and, by June 19, 2021, a total of 831,213 vaccine doses had been administered. Uganda experienced a second wave of COVID-19. At the beginning of the second wave, April 2021, cumulative numbers of COVID-19 cases for Uganda were 44,594 with 361 lives lost. Numbers of cases and deaths rose rapidly in the second wave with a shorter span of 10 days to reach full-blown pandemic (Ministry of Health, 2021).

A case study of Nakawa Division, one of the five divisions of Kampala Capital City Authority (KCCA), reflected a bioethical problem of noncompliance to standard operating procedures (SOPs) as a public health measure to prevent COVID-19. Kampala is the main city of Uganda. The country runs a multiparty political system of governance. According to Uganda Bureau of Statistics (UBOS, 2020), country's population is 41 million with a young population (under 18) projected at 53%; older persons (aged 60 years+) form 4% while urban population constitutes 27% of population. With an estimated 77% of youth under 30 years, Uganda has the youngest age structure in the world. Distribution of Uganda's wealth is skewed with 20.3% of population living under the poverty line. Literacy rate is 76%. The working population is approximately 15 million of which 6 million are engaged in subsistence agriculture and a sizeable proportion involved in informal sector. Kampala city holds 4.3% of the working population.

This literary enquiry aims at evaluating adherence to COVID-19 containment SOPs in cosmopolitan Kampala, Uganda, as this impacts the spread of the disease and related deaths. Findings will be documented to contribute to the body of knowledge to inform individuals, the public, researchers, and policymakers. An interdisciplinary approach was applied in analyzing adherence to SOPs. This approach embraces a range of disciplines such as medicine, law, politics, and community development. The approach underscores that COVID-19 management is complex and multidimensional, therefore begging the involvement of stakeholders from diverse disciplines. The chapter benefits from references to the pre-COVID-19 dynamics of public health adherence to prevent the spread of outbreaks and ethical challenges experienced during Ebola outbreak in Ugandan and West Africa.

**Goal and Objectives of the Study**
**Goal**

To explore individual and community adherence to SOPs as a public health measure to prevent COVID-19

**Objectives**

1. To establish if individuals properly wear their masks in public spaces
2. To determine if individuals practice proper handwashing in public places where handwashing facilities are provided
3. To assess if individuals observe social distancing measures while in public places
4. To explore human behavior factors associated with nonadherence to COVID-19 SOPs and responsive strategies

## 2 Statement of the Problem

COVID-19 pandemic in Uganda worsened in the second wave with a sevenfold increase in daily number of new cases and upsurge in deaths. Country registered highest number of cases in a single day in June 2020 with 1259 cases out of 7424 tests representing a 17% positivity rate (GOU June 2021b). Kampala reported highest case load averaging 500+ cases per day (GOU, June 2021b). Whereas elderly segment (45 years+) were most affected during the first wave, records showed a switch with youth 20–39 years most affected in the second wave and increasing transmission among 10–19-year-old as well as escalation in severity of disease (Ministry of Health, 2021). GOU adopted SOPs/MHS as a policy and public health measure to effectively prevent transmission of COVID-19. Bwambale (2021) reports GOU adopted 33+ WHO-recommended measures including social distancing, handwashing, and wearing face mask, which formed part of preparedness plans in COVID-19 prevention. Relatedly, Vladimir Cvetković et al. (n.d.) advise that in the absence of vaccines or medications, nonpharmaceutical interventions remain an outstanding option. In resonance, Manikandan (2020) reechoes aforementioned while also concerned that maintaining social distancing (distancing 6 ft/or 2 m) in public spaces is difficult to achieve. Manikandan notes that Centres for Disease Control and Prevention recommends that wearing of cloth-based face masks in public spaces prevents risk of SARS-CoV-2 transmission, yet information on adherence to this and other aforementioned SOPs remained scanty.

GOU cautioned on poor compliance, so did Ssejjoba's (2020) report. Additionally, Makerere University study alluded to compliance gaps—that respondents aged 50+ years had a 26% higher adherence to COVID-19 guidelines compared to younger persons below 25 years. According to GOU, situation was aggravated by entry of new variants, Indian, South African, and UK variants, and their complexities, aggression, sustained transmission, and shift of age group from older to younger.

Spaces which called for urgent individual vigilance and responsible COVID-19 prevention compliant behavior included crowded market places, public transport parking yards, and banking halls. Relatedly, congestion in refugee setting was noted as a feasibility challenge in this location (Ssejjoba, 2020). Informal communication suggested wide nonadherence to masks and revealed excuses for nonuse of masks including discomfort. Additionally, inappropriate mask use, mask sharing, or wearing mask simply to avoid police confrontation were exposed by Ssejoba.

Duty bearer concern on mask use continued. In a May 2021 television broadcast of NTV channel, Uganda's Ministry of Health Permanent Secretary, Dr Atwiine, was screened urging for a lockdown since Ugandans were disregarding COVID-19 prevention measures. Dr. Atwiine urged citizens "*If Covid is changing daily to survive why can't we adjust to survive...*" . [Tweet]. Atwiine, D[@DianaAtwiine] (2021, May 27). The committee assessing progress of preventing Corona established nonadherence to SOPs (Buyondo, 2020). Nonadherence to SOPs violates moral and ethical principles, negates values, and presents a notable bioethical problem threatening human health and life. The vice is a violation of both the God given life gift and rights enshrined in Universal Declaration of Human Rights (UDHR, 1948) and guidelines of Universal Declaration on Bioethics and Human Rights (UDBHR, 2005). SOP infringement represents ethical violation of beneficence and nonmaleficence principles. At the same time, a reflection on and analysis of SOPs advocacy, uptake, and adherence raises questions of autonomy and individual choice vis a vis risk to community and fair play, reflective of a moral and ethical dilemma discussed further below.

Variables such as character, culture, and demographics (age dynamics, education, socioeconomic background) invariably influence human behaviors, consequently affecting adherence to SOPs. Cultural practices, values, norms, behaviors, and experiences are associated with or result into adherence/nonadherence to SOPs and play a crucial part in exposing persons to risk of disease. Bwambale mentions that 21.7% respondents indicated they did not wear a mask in public places as required, while Finset et al. (2020) describe awkwardness and discomfort associated with prescribed COVID-19 prevention compliant behavior.

Cultures, such as the hug greeting practiced in Western Uganda, conflicted with and, according to Sturm and Prudden (2019), and Ssejjoba (2020), negated

COVID-19 prevention efforts of social distancing, hence created an ethical dilemma. Ssejoba further noted that discouraging cultural practices such as eating from same plate proved difficult. Similarly, wedding celebrations, spiritual/religious activities, and burial ceremonies have been a way of life pre-COVID-19, which attract big crowds and are risky. Mentioned activities were banned in Uganda in early 2020 and replaced with scientific functions (weddings, prayers; https://religionunplugged.com, July 2020), restricting numbers to only ten attendees. While this measure was intended to balance values and COVID-19 prevention principles, it conflicted the autonomy principle. When government relaxed above mentioned ban and allowed reopening of churches and wedding functions but with strict restricted numbers and SOP adherence, communities contravened restrictions and instead organized festivities hosting 200+ people. Burial cultural practices and associated huge numbers were another source of concern highlighted by office of the president on April 18, 2021, asking radio stations to withhold details of funerals so that communities could not convene for funeral ceremonies. Nevertheless, this action was unfair and infringed self-determination and autonomy principle. Aforedescribed scenario exemplified ethical dilemmas and role of a risk/benefit analysis and balances in decision-making. Issue of ethical dilemma further surfaced in reports from Nakasongola District wherein SOP nonadherence was addressed by armed personnel caning noncompliant individuals, causing bodily harm, violating autonomy, and human rights. Meanwhile, duty bearers continuously cautioned its citizenry. Irene Nakasiita of Uganda Red Cross Society stressed, "You die alone and that's why we call upon Ugandans to embrace the SOPs, to keep safe" (Buyondo, 2020).

Handwashing advocacy has been ongoing for many decades to date in Uganda and elsewhere. Reports, for instance, Hirai (2016), highlight poor handwashing habits of individuals and communities in Uganda. Culturally, most communities in Uganda are used to irregular practice of washing hands only after eating. In resonance, Hirai describes washing hands after eating and after cleaning child stools as key handwashing moments for communities in Indonesia and sub-Saharan African countries, implying communities consider handwashing a challenge. According to Bwambale, 10% of respondents in Makerere University study did not wash their hands with soap and water or sanitize.

**Lockdown** instituted by most countries typified social distancing as a measure to prevent COVID-19 spread and/or flatten the curve. Reports indicated high levels of resistance to lockdown globally, hence a risk to health and life, contradicting Universal Declaration of Human Rights. Bwambale reported 27.7% respondents did not observe social distancing of 2 m.

For any new disease, communities need information in order to make personal choices to protect themselves and others. Comprehensive, clear, understandable information and strategic communication to communities about a disease of the magnitude of COVID-19 was an imperative. Finset et al. (2020) underscore communication as key in promoting behavior change. However, communities, communication experts, and critiques expressed concern of huge communication and information gaps amid widespread myths. Bwambale (i) reported that 26% of respondents in a study conducted by Makerere University contended there is no

COVID-19 in Uganda; (ii) found dangerous level of myths and misconceptions, 17.8% respondents believed herbal medicines eliminated risk of COVID-19 infection, 32.7% believed prior exposure to the disease was protective, hence eliminated need to observe SOPs, and 15.4% believed disease was for foreigners and travelers from abroad. Similarly, Ssejjoba (2020) reported belief in myths and disbelief in seriousness of Corona among refugee communities. A parliament of Uganda circular of June 23, 2021, testified on rumor mongering, which put the number of members of parliament infected with coronavirus to 200 against correct number of 14. Social media platforms, especially for urban setting, were key in rapid and massive spread of myths and misconceptions. Bwambale noted social media was among popular sources of information on COVID-19 at 23%, with television (75.4%) and radio (67.2%). Meanwhile, critiques maintained that duty bearers did not facilitate process of engaging communities and stakeholders to produce simplified, comprehensive information on COVID-19 and that top-down information, education, communication (IEC) approaches, and associated threat messages were grossly counterproductive.

Finset et al. (2020) highlighted challenge of transition from knowing to doing and the intention-behavior gap. Many Ugandans demonstrated negative attitudes toward MHS. Furthermore, aspects of globalization, changing lifestyle, and breakdown of positive cultural norms are fertile breeding ground for moral decadence. This situation is associated with loosening of solidarity, Ubuntu, and truth-telling ethical principles. Uganda police force official shared case of individuals presenting fake COVID-19 positive medical certificates to enable them drive during lockdown period. "*Our officers quickly wave them off than risk "catching" it*". [Tweet]. Kasingye, A[@AsanKasingye] (2021, June 23).

Amid huge COVID-19 information bases and reports about the disease (Bwambale, 2021; Ssejjoba, 2020), literature pinpointing adherence to SOPs as a public health measure in prevention of spread of the disease in Uganda remains scanty as re-echoed by Matovu et al. (2021).

## 3 Methodology: Document Review

**Documentary research** was conducted by reviewing electronic documents and publications including journals, E-books, theses, conference, and meeting proceedings, complemented by social media, electronic, and print media and reports. One hundred journals were reviewed by the researcher, guided by study objectives, authors' experience with COVID-19, and capturing most recent information covering a range of disciplines including medical, political, and legal. Data was analyzed and synthesized by way of interdisciplinary approaches, tracking recurrent and common themes.

**Delimitations, Scope of the Study, and Study Area:** Kampala Capital City Authority (KCCA) was purposively selected for this study to align with principles of cost effectiveness and efficiency. KCCA is quite representative of Uganda and has urban, periurban, and semblance of rural settings.

**Study Variables—Independent Variables**: Age, sex, education, marital status, socioeconomic status, employment, workplace/environment, and residency (semirural, urban, periurban). **Dependent Variables:** SOP behavior.

**Introduction**

Numerous strategies were devised by governments to counteract COVID-19. Prevention measures were tied up with behavior change and its complexities; hence, continuous studies were necessary to inform interventions for the complex COVID-19 disease. Nonadherence to SOPs posed numerous questions that cut across various disciplines, underscoring interdisciplinarity. Finset et al. (2020) assert that studies of previous pandemics demonstrate that official recommendations and policies are met with skepticism relating to noncompliance to SOPs. This exploratory study was largely confined to behavioral aspects of management of COVID-19. The literature review covered demographic factors, socioeconomic aspects, lack of supplies, adherence to SOPs, COVID-19-related IEC, and behavior change.

**Demographic factors:** age, education, gender, marital status, and socioeconomic elements were likely to impact the uptake of SOPs.

## *3.1 Adherence and Nonadherence to SOPs*

Matovu et al. (2021) highlighted growing evidence of challenges with adherence to COVID-19 SOPs and decried related limited documentation in sub-Saharan Africa; this upheld the need of SOP adherence knowledge.

**Handwashing Practice**

Pre-COVID-19 handwashing practice has a bearing on uptake and adherence to handwashing as part of the COVID-19 preventive measures package. Policymakers and programmers needed to tap into body of knowledge on handwashing pre-COVID-19 when devising strategies to scale up handwashing practice. Hirai (2016) reports interpersonal influences, presence of handwashing facility, and key handwashing moments (after eating and after cleaning child stools) among key potential determinants of handwashing with soap. This finding ropes in innovative descriptions such as "key handwashing moments" and informs future IEC strategy redevelopment. Hirai further notes the positive effect of availability of supplies mentioned above. In resonance, the Ebola management experience in West Africa highlights availability of supplies as a facilitator of SOP adherence (Kerstiens & Matthys, 1999; Muyembe-Tamfum et al., 1999).

A lack of supplies in a context of deepened poverty among the Ugandan population undermines their right to health and life and the application of the ethical principle of justice and is a constrainer to public health interventions in emergency situations. Access to resources and supplies for vulnerable communities remains a responsibility of government. For instance, during the 2000/2001 Ebola outbreak in Gulu, northern Uganda, the inadequate quality of protective materials was a challenge and contributed to transmission of the Ebola virus within the healthcare setting. Meanwhile, for the Ebola outbreak in Bundibugyo, western Uganda, most transmission was associated with handling of dead persons without appropriate protection Wamala et al. (2010).

**Social Distancing**
Ssematimba et al. (n.d.), Ngonghala et al. (n.d.-b), and other authors strongly advocate the use of masks and social distancing combination as safeguards in preventing COVID-19.

## 3.2 *COVID-19 Prevention IEC and Behavior Change Communication (BCC) and Effects of Top-Down Strategies*

Individuals and community possibly were not adhering to SOPs in a show of resistance to pressure. Public health definition and behavior change principles are not aligned with pressure. Perhaps SOPs were not perceived as their "informed choice" but rather a behavior being forced onto them in a top-down manner. Awobamise et al. (2021)'s evaluation of Uganda Government communication strategy echoed this noninvolvement of communities in communication development strategies for a matter (COVID-19), which directly and hugely affected communities. Relatedly, Manikandan (2020) contended information dissemination about COVID-19 motivated individual to embrace prevention measures, SOPs, as opposed to government-initiated interventions such as lockdown or even penalties. The latter may have sacrificed autonomy but may have served as a public good, perpetuated utilitarian benefits and resulted in reduced effects of COVID-19. Further to the above, in their elaboration, Monagle and Thomasma (2005) assert that the practice of informed consent hinged on the respect for autonomy principle is imperative to recognize and promote the autonomous actions of the client. This also involves a communication process that is compassionate and respectful to the client's unique values, even if they differ from the standard goals of biomedicine. Mindfulness of illiteracy challenges is important—(Doe-Anderson et al., 2016; Schieffelin et al., 2016) highlight the negative impact of low literacy on Ebola activities (studies, informed consenting). Finset et al. (2020) emphasize the need to provide clear, honest, and valid information to the public, with a conclusion of verified information as most effective against COVID-19. The study by Bwambale (2021) highlights that 99.8% respondents had heard about COVID-19 and its spread mechanism, a finding not ruling out knowledge gaps on other aspects relating to new disease. Interestingly, findings by Bwambale revealed trust in Uganda's health workers regarding knowledge of science and capacity to control COVID-19. Similarly, application of the principle of justice contributed to effective community mobilization, which was critical in controlling Ebola epidemics through regular provision of information to the public and minimizing of rumors and misconceptions about the disease/public health emergency (Okware et al., 2022).

### *3.3 Behavior Change*

Role of behavior change is overbearing in the development field and is a cornerstone in adherence to SOPs and urgent prevention of COVID-19. Indeed, as Michie et al. (2021) reported, behavior contributes hugely to mortality and morbidity and this dictated urgent action guided by principles such as "Do unto others as you would have them do unto you." May (n.d.). While critiques have expressed misgiving on behavior change strategy, programmatic evaluations, behavioral experts, and studies demonstrated phenomenal achievements of behavior change. Michie et al. (2021) underscored emphasis and prospects of applying theory to develop and interpret success or otherwise of behavior change interventions. Remarkable finding recaptured optimism in pursuing behavior change.

## 4 Discussion: Interdisciplinary Diagnosis and Approach

COVID-19 preventive measures (SOPs) were instituted at the global and country levels. This study explored **bioethical problem of nonadherence to SOPs** as a public health measure to prevent transmission of coronavirus amid escalation of COVID-19. Data was collected using documentary research methodology and enriched with most current views of government leaders.

The magnitude of the problem was reflected in statistical presentations and analyses provided overtime. Additionally, dimensions of bioethics problem included historical perspective, law, scientific aspects, medicine, technology, socioeconomics, politics/political science, policy, anthropology, sociology, psychology, culture, ethnicity, theology, demographics, human behavior, environment, IEC, community development, and mobilization. Indeed, in a video (Columbia, 2016), it was emphasized that health care did not only cover science. Bioethical problem attracted a review of theories and called for a recheck of possible conflicts and compromise to key principles, beneficence, maleficence, justice, and autonomy. Complexity of the problem begged understanding and inputs from a range of perspectives. It is multidimensional and concerns a person, described by Zasukhina (2014), as an organic whole. Interdisciplinary approach guided discussion on action of uptake and adherence to SOPs in the prevention of COVID-19. Described as a methodology to bioethics wherein knowledge is found in and drawn from more than one discipline, Franciscan University (2014) avers the approach attempts to answer question "what should be done?" As Franciscan University suggests, this approach offers numerous advantages in analyzing, diagnosing, and managing the bioethical problem. Relatedly, Segalàs et al. (2014) suggest interdisciplinarity mobilizes two or more disciplines for one activity—disciplines (elaborated in the above second paragraph) speak to each other in the process. Formation of Internal Review Boards recognizes interdisciplinarity in solving bioethical issues. Interdisciplinary diagnosis highlighting problem, interdisciplinary perspective, variables at stake, agents, precepts, principles observed or violated, values at play, moral challenges in acts, and circumstances of bioethical case are discussed below.

**History** Modernity is enriched by historical experiences. Uganda has numerous historical examples of managing challenging epidemics. Lessons may be picked from such records including (i). Bubonic plague (1910–1946)—managed by motivating citizens to kill the rats through a payment for rat tails which were presented to authority (**utilitarian approach, benefits to the actors**). (ii). Citizens were forced (caned "kibooko") to grow coffee (**common good achieved, top/down approach, infringed on autonomy, human rights, happiness**). (iii). HIV prevention management (in the early 1980s)—IEC-discouraged fear messages (iv). Whether or not to expend efforts on pinpointing the origin of COVID-19. Heated debate on BBC and other media blaming China for manufacturing coronavirus raised many issues. History informs us that a similar process of establishing the origin of HIV was undertaken in the early 1980s. Historical angle plays a fundamental role in processing bioethical issues.

**Scientific Knowledge and Statistics Is of Significance in COVID-19** Cause of COVID-19 was defined—SARS-CoV-2. This and the mode of transmission of the virus, its fragility, formed the scientific basis of SOPs in the prevention of coronavirus spread. Yet Hadorn et al. (2008) reckoned that statistics does not lead to clear-cut answers for complex matters, in this case adherence to SOPs.

Scientific evidence established SOPs in the prevention of COVID-19 spread. Allied with this, researchers such as Cvetković et al. (n.d.) and Manikandan (2020) endorsed mask use, handwashing, and sanitizer use (MHS) with Manikandan, however, cautioning that observing social distancing in public spaces may be challenging to achieve. This argument aligns with socioeconomic barriers to up take and adherence to MHS elaborated further in the paragraph below (socioeconomic status of population, etc.). Seventy percent of households in Kampala use only one room for sleeping (Hashemi et al., 2015), with 3+ people crammed in that small room for 7+ h. This defeated social distancing measure. Further, the range of masks and handwashing facilities on market questions on their safety and prices, raised issues of corruption, injustices, and controversy. This development (i) further constrained uptake of SOPs and (ii) attracted need for recruiting additional disciplines to interrogate on newly introduced issues (safety and exploitation); hence, economists, rights advocates, health and safety, and biochemists were invited on board in addition to scientists and city planners.

Socioeconomic status of population stood out in SOP adherence. Majority of Kampala city residents/workers are youthful, urban poor, engaged in petty trade, earn daily, and live hand to mouth. The typical workplace, down town Kampala in Kikuubo trading complex, several arcades, and market places are always overcrowded, wherein observing social distance was simply an illusion. Obtaining equilibrium was among the challenges facing bioethicist—sustain livelihoods while observing SOPs.

A dialogue was long overdue; government, city and national planners, sociologists, and human rights advocates needed to address this hopeless situation—distributive justice gaps/inequity in distribution of resources (abject poverty). Government bears responsibility for well-being of all its citizens and value of life of all human beings regardless of their socioeconomic status is irreducible. Rights of impoverished Kampala city residents to health, life, and decent employment enshrined in Universal Declaration of Human Rights and also in the country constitution had to be upheld. The IEC dialogue in the second last paragraph below is closely associated with socioeconomic debate.

Alongside all above, politics, policy, law, anthropology, culture, ethnicity, theology, and demographics (gender, age dynamics, education) were demonstrated to impact SOP uptake.

Relatedly, numerous voices advocated separating COVID-19 disease from politics. Disease touched spiritual, physical, emotional, psychological, and socioeconomic aspects of life, attracted huge financial resources, raised cross-border questions, and beared outright policy and legal implications. Call for personal responsibility to prevent disease pointed to a stake holding relationship. Failure to fulfill this responsibility due to anthropological reasons was met with new policies, directives, or laws as evidenced in June 18, 2021, presidential address/directives on COVID-19 resurgence, which provided for more stringent measures: (i) longer curfew period, (ii) places of worship remained suspended (theology), and (iii) burials (cultural and ethnicity aspect) restricted to 20 people. Resident District Commissioner's communication to radio stations to desist from announcing dates of funerals in order to abide with presidential directive of limiting crowding, highlighted interdisciplinarity in solving bioethical dilemmas (Office of President, 18th April 2021). Policies arising of political directives created/strengthened roles of agents in the interdisciplinary approach; for instance, President directed Ministry of Health to strengthen the village health teams in the COVID-19 management. Directives (and laws) indeed are enforceable (Segalàs et al., 2014) without regard to individual's conscience. Further, rallying various disciplines to review progress of projects and initiatives such as SOP uptake, jointly review targets, arguably creates mutual trust as Hirsch et al. relate and avails opportunity for building broad common knowledge base about initiative. Demographics (age, gender mainstreaming): age: country experienced a shift in vulnerability from elderly to youth with second wave. Huge numbers of youth increasingly crowded vaccination centers, which they earlier had shunned presumably due to high risk perception. Gender mainstreaming: Less SOP compliance among men deserved further interrogation.

Human behavior (aspect of anthropology) is allied to scientific knowledge in efforts to prevent disease. Violation of SOP guidelines (i) is out rightly bad, wrong, inherently immoral, and contravenes deontology ethical theory and (ii) is a breach of duty and responsibility on part of public and translates into risk of spread of disease and breaches utilitarianism and common good. Neuroethics discipline and behavior genetics elaborate role of genetic and environmental influences on human behavior. Life, our gift from God, pronounces wellness, happiness, and sustainability. The narrative, amid a worsening COVID-19 situation, clearly pointed to huge gaps of nonadherence to SOPs. Permanent Secretary for health, Dr Atwiine, alluded to this nonadherence in statement that adherence to SOPs to dot was a must or else we perish (Ministry of Health, 2021). Nonadherence thence needed to be further understood and managed. Adjustment to new normal (a divergence) was destined to become the norm. However, it was noteworthy from communication experts that fear messages (do this or "perish") were counterproductive. Trust, freedom, dignity, equity, and mercifulness were beckoning in behavior change strategy.

Furthermore, diseases such as Corona were linked with migration of animals and associated with environmental degradation. Approaches to mitigate environmental degradation, a sensitive issue of global scale, included global guidelines such as the Kyoto protocol as well as local guidelines, rules, and laws. Social media pollution dimension is another ethical challenge of modern society, which heavily contributed to myths and misinformation on COVID-19 prevention SOPs (discussed further in 9.8). Experts guided that a control measure to manage social media pollution ought to embody the utilitarian value and not simply be authoritarian and/or for benefit of duty bearer. A harmonized view of environmentalists, scientists, ICT experts, and sociologists was an eye opener in seeking a solution for such above complex bioethics issues.

Documentation revealed several challenges in pursuance of the interdisciplinary approach.

Integration of disciplines per se is hampered by "not speaking the same language" according to Hadorn et al. (2008). This constraint is extendable to IEC, a tool in addressing adherence to SOPs. Comprehensive, simple, clear, understandable information, and strategic communication to communities about COVID-19 is an imperative. Finset et al. (2020) and other actors reecho this while expressing urgent concern of huge communication and information gaps amid wide spread myths. Bwambale (2021) reports that a sizeable 26% of respondents in the study conducted by Makerere University contend there is no COVID-19 in Uganda. Study also established dangerous level of myths and misconceptions; for instance, 17.8% respondents believed users of herbal medicines are not at risk of COVID-19 infection while 32.7% believed a prior exposure to the disease is protective; hence, those already exposed may not need to observe SOPs. Similarly, Ssejjoba (2020) reports myths and disbelief in seriousness of Corona among refugee communities. With such myths/misinformation deterring people from adopting SOPs, the urgency for clear, understandable information, and communication was great. Additionally, strategy to engage communities and other stakeholders to produce unified, simplified, comprehensive COVID-19 information was greatly important. In furtherance of this, Hirsch et al., recommend fostering mutual understanding through effective communication, use of everyday language, understanding of specialized definitions, and discipline terms and integrating stakeholders. Comparatively, in the EVD epidemic response, panic and confusion arose due to beliefs of the causes of EVD, including those who believed national or international governments had infected

West Africans with Ebola for their gain. Ethical challenges also arose due to fear, violence, community mistrust, and failures for health workers to communicate adequately with the frightened communities, which further complicated the Ebola treatment and response. This slowed down the effectiveness of the public health interventions and tampered with application of the ethical principles of nonmaleficence and beneficence. Furthermore, authoritarian, top-down approaches in IEC are alien to interdisciplinary spirit. Worse still, authoritarian approaches usually carry threat messages which are grossly counterproductive and untenable. History of such approaches with HIV prevention proves method unfruitful, while experience shows nonparticipation results in communities ignoring the measures (Hirsch et al. 2008) and products. Further, the cultural perspective becomes useful in IEC, whereby the use of idioms, folklore, proverbs, and familiar stories makes message relatable and understandable. In addition, urgency to deal with rumors, myths, misconceptions, and disinformation rife in COVID-19, especially in light of social media platform activities (highlighted by Bwambale), was glaring.

In recognition of the saying, "nothing for us without us," community development and mobilization discipline remained a cornerstone in driving adherence to SOPs. The power of engaging communities as key agents in solving their problems was a winning strategy. Hadorn et al. (2008) argued that community contribution went far beyond putting critical questions to expert and that in their creative engagement, community asked provocative, mind-boggling open-ended questions. Community participation developed and reinforced a sense of ownership of the problem crucial for their motivation. Further, community are stimulated to deploy their "extended facts," including anecdotal evidence, confidential information, personal testimony, local knowledge, and ethical commitments, all of which translate into adoption of desired behavior (SOPs) and adherence. Besides, as reflected by Hadorn et al., opportunity shows for home grown local experts and practitioners to fill in with their experience based knowledge, resulting in robust solutions. Internal Review Boards were examples of systems aligned with notion of interdisciplinarity whose benefits were clear.

## 5 Conclusion

The Alma-Ata (1978) stresses the right and duty of individuals to participate in planning and promoting their health. COVID-19 pandemic highlighted role of public health in protecting life and health, importance of personal responsibility in public health, well-being of communities, nations, and the globe.

Study findings revealed alarming levels of nonadherence to SOPs while emphasizing adherence to SOPs as the silver bullet. Appropriate behavior of adherence to SOPs was constrained by demographic, socioeconomic, legal, political, and a host of other factors cutting across numerous disciplines—a complex of things.

Demonstrated complexity of behavior change and associated factors called for ground breaking measures and interventions espousing the interdisciplinary, multidimensional, and teamwork approach. Capitalizing on the multidisciplinarity bioethical approach and its application of ethical reasoning according to Franciscan University (2014) is the way to go in solving SOP adherence challenges. Changing behavior and sustaining new behavior both take up numerous dimensions, factors, and processes. Principles and values have been shown to guide behavior change process, while decisions entail balancing elements at different levels. Sometimes, principles and values got compromised; for instance, cultural values (for burial ceremonies) were rested to facilitate adherence to SOPs. It is anticipated that this study will be a springboard for stronger SOP adherence mechanisms, just as experiences such as the Tuskegee study sparked off initiatives including the signing of the National Research Act into law (1974) and creation of the National Commission for Protection of Human Subjects of Biomedical and Behavioral Research.

# 6 Recommendations

**Addressing Moral Wrongs** Within interdisciplinary approach, the nation, alongside its country motto, "For God and my country" ought to tactfully establish mechanisms to reawaken citizenry to reality, sacrosanctity, and protectiveness of this motto, which also is the epitome of the common good principle. Immoral acts exposed in this study were a reflection of derailment from country motto and abandonment of religion. Tham (2007) cautions about marginalization and secularization of religion (Chapter 2, page 110–168) and suggest how to rejuvenate role of religion in bioethics (Chapter 6, page 466–487) for a worthwhile life experience. Further, stakeholders need to foster and redesign existing ethics and moral reawakening programs such as "Patriotism in the Training of Early Childhood Teachers in Uganda (PATI Project)" alongside global initiatives, for instance, workable ethics and compliance program options suggested by Chilima and Lamont (2018).[1]

Duty bearers with support from relevant disciplines [communication, history, political science, sociology (human rights, anthropology)] should take practical steps to champion and consolidate trust (show are in tune with reality), just as video[2] demonstrated.

[1] Is State Vice President of the Republic of Malawi (2014–2019) and UTM Party President and Presidential candidate for elections due in May 2019. He is a results-oriented and people-centric leader. He holds a Ph.D. in Knowledge Management from the University of Bolton (UK), as well as a Master's degree in Economics and a Bachelor's degree, both from the University of Malawi. He is also accredited by Louis-Allen Worldwide as a Leadership Development Trainer (Saulos Klaus Chilima, Bruce Lamont, May 2018.)

[2] Talks at Columbia (2016)—a doctor experiences what it is like to be a patient; the worries and emotions awaken him to be compassionate.

**Alternative Measures: Vaccination** It is imperative to implement scientifically proven and viable measures such as vaccination to address complexity of observing SOPs in crowded workspaces such as Kikuubo[3] trading center as a priority.

**IEC/BCC** Deepen engagement with stakeholders to jointly identify IEC/BCC gaps and responsive measures to drive uptake of SOPs.

**Laws, Policy, and Guidelines** Interdisciplinarity approach and best practices of developing policies, guidelines, and laws should not be sacrificed at any time as so doing is likely to do more harm.

**Socioeconomic Aspects** Financial and physical access to quality products are imperative to facilitate the uptake of SOPs.

Further research is recommended, interrogating with communities in Kampala and its strategic position as capital city of Uganda, main business hub as well as administrative seat. Future research on adherence to SOPs should consider interdisciplinarity and follow-up action should reengage action-oriented representatives from various disciplines to inform subsequent strategies, devoid of abstract, and "vague and meaningless" approaches, denounced by Segalàs et al. (2014). Development-focused funding agencies are invited to invest in interdisciplinarity and capitalize on renewed thinking of multiskilled, multifaceted research/implementation.

## References

Alma-Ata. (1978). *Primary health care. Report of the international conference on primary health care. Alma-Ata, USSR, 6–12 September 1978*. World Health Organization.

Atwiine, D. [@DianaAtwiine]. (2021, May 27).

Awobamise, A. O., Jarrar, Y., & Okiyi, G. (2021). Evaluation of the Ugandan government's communication strategies of the COVID-19 pandemic. *Online Journal of Communication and Media Technologies, 11*(1), e2021xx. https://doi.org/10.30935/ojcmt/10824

Buyondo, D. (2020, November 9). Complacency hampers COVID-19 fight. *New Vision*, p. 16.

Bwambale, T. (2021). 26% of Ugandans believe there is no COVID-19 in Uganda-study. *New Vision.*

Chilima, S. K., & Lamont, B. (2018, May). Addressing moral decadence: Towards an effective compliance and ethics program. *Africa Journal of Management*. https://doi.org/10.1080/23322373.2019.1565063. https://www.tandfonline.com/doi/abs/10.1080/23322373.2019.1565063. Accessed on 25 June 2021. Google Scholar.

Columbia, D. (2016). *Formulating a research problem*. [Video file]. Retrieved May 28, 2019, from https://www.youtube.com/watch?v=KywZXYXiQ8s

Cucinotta, D., & Vanelli, M. (2020). WHO declares COVID-19 pandemic. PMID: 32191675. PMCID: PMC7569573. https://doi.org/10.23750/abm.v91i1.9397. Free PMC article. https://pubmed.n.cbi.nlm.nih.gov/32191675/. Accessed on 29 May 2021.

[3] Kikuubo trading center in down town Kampala is a highly congested place with retail and wholesale traders dealing in mixed merchandise; traders travel from all over the country.

Cvetković, V. M., Nikolić, N., Radovanović, U., Öcal, N. A., Noji, E. K., & Zečević, M. (n.d.). Preparedness and preventive behaviors for a pandemic disaster caused by COVID-19 in Serbia. PMID: 32527056. PMCID: PMC7313005. https://doi.org/10.3390/ijerph17114124. Free PMC article. https://pubmed.ncbi.nlm.nih.gov/32527056/. Accessed 13 June 2021. On Pubmed.

Doe-Anderson, J., et al. (2016). Beating the odds: successful establishment of a phase II/III clinical research trial in resource-poor Liberia during the largest-ever Ebola outbreak. *Contemporary Clinical Trials Communications, 4*, 68–73.

Finset, A., Bosworth, H., Butow, P., Gulbrandsen, P., Hulsman, R. L., Pieterse, A. H., Street, R., Tschoetschel, R., & van Weert, J. (2020, May). Effective health communication—A key factor in fighting the COVID-19 pandemic. *Patient Education and Counseling, 103*(5), 873–876. Published online 2020 April 23. https://doi.org/10.1016/j.pec.2020.03.027. PMCID: PMC7180027. PMID: 32336348. Elsevier Public Health Emergency Collection. https://www.ncbi.nlm.nih.gov/pmc/articles/PMC7180027/. Accessed on 30 May 2021.

Franciscan University of Steubenville (Producer). (2014, July 2). *Dr. Christian Brugger: Bioethics as interdisciplinary*. [Video file]. Retrieved June 19, 2019, from https://www.youtube.com/watch?v=6%2D%2D-q9NILH7ls

Government of Uganda (GOU). (2021a, June). Address by H. E. Yoweri Kaguta Museveni President of the Republic of Uganda to the nation on COVID-19 pandemic resurgence as of 18th June, 2021, Nakasero.

Government of Uganda (GOU). (2021b, June). Address by H.E. Yoweri Kaguta Museveni President of the Republic of Uganda to the Nation on COVID-19 resurgence and current status of the pandemic in the country 6 June 2021, Nakasero.

Hadorn, H., Hoffmann, G., Riem, H., Klemm, B., Grossenbacher, S., Mansuy, W., Joye, D., Pohl, C., & Zemp, E. (Eds.). (2008). *Handbook of transdisciplinary research*. Springer.

Hashemi, A., Cruickshank, H., & Cheshmehzangi, A. (2015). *Environmental impacts and embodied energy of construction methods and materials in low-income tropical housing*. https://www.mdpi.com/2071-1050/7/6/7866/htm. Accessed on 18 June 2021.

Hirai, M. (2016). Water, Sanitation, and Hygiene (WASH) for Child Health:Evidence from Large Household Surveys in Indonesia and Sub-Saharan Africa. Accessed June 6th 2021 on https://www.proquest.com/openview/eb248137ba4e5128417114c3317dcf75/1?pq-origsite=gscholar&cbl=18750&diss=y; Google Scholar.

Hirsch, H, Hoffmann—Riem, H., Biber—Klemm, S., Grossenbacher—Mansuy, W., Joye, D., Pohl, C.,... Zemp, E. (Eds.). (2008). *Handbook of Transdisciplinary Research*. Berlin: Springer.

Kasingye, A. [@AsanKasingye]. (2021, June 23).

Kerstiens, B., & Matthys, F. (1999). Interventions to control virus transmission during an outbreak of Ebola Haemorrhagic fever: Experiences from Kikwit, Democratic Republic of the Congo 1995. *The Journal of Infectious Diseases, 179*(supplement_1), S263–S267.

Manikandan, N. (2020, July–December). Are social distancing, hand washing and wearing masks appropriate measures to mitigate transmission of COVID-19? *Vacunas, 21*(2), 136–137.

Matovu, J. K. B., Kabwama, S. N., Ssekamatte, T., & Ssenkusu, J., Wanyenze, R.K. (2021). COVID-19 awareness, adoption of COVID-19 preventive measures, and effects of COVID-19 lockdown among adolescent boys and young men in Kampala, Uganda. *Journal of Community Health*. https://doi.org/10.1007/s10900-021-00961-w. Accepted: 4 January 2021 ORIGINAL PAPER © The Author(s), under exclusive licence to Springer Science+Business Media, LLC part of Springer Nature 2021. Accessed on 4 June 2021 on Google Scholar.

May, D. R. (n.d.). Steps of the ethical decision—Making process. Retrieved May 27, 2019, from https://research.ku.edu/sites/research.ku.edu/files/docs/EESE_EthicalDecisionmakingFramework.pdfo

Michie, S., Johnston, M., Rothman, A. J., de Bruin, M., Kelly, M. P., Carey, R. N., et al. (2021). Developing an evidence-based online method of linking behaviour change techniques and theoretical mechanisms of action: A multiple methods study. *Health Services and Delivery Research, 9*(1) https://www.ncbi.nlm.nih.gov/books/NBK567047/pdf/Bookshelf_NBK567047.pdf. Accessed on 6 June 2021. On Medline

Ministry of Health. (2021). Working together to stop the spread of COVID-19, Sunday 20 June 2021.

Monagle, J. F., & Thomasma, D. C. (2005). *Health care ethics: Critical issues for the 21st century*. Jones & Bartlett Learning.

Muyembe-Tamfum, J., et al. (1999). Ebola outbreak in Kikwit, Democratic Republic of the Congo: Discovery and control measures. *The Journal of Infectious Diseases, 179*(Supplement-1), S259–S262.

Ngonghala, C. N., Iboi, E., Eikenberry, S., Scotch, M., MacIntyre, C. R., Bonds, M. H., & Gumel, A. B. (n.d.-a). Mathematical assessment of the impact of non-pharmaceutical interventions on curtailing the 2019 novel Coronavirus. https://pubmed.ncbi.nlm.nih.gov/?term=ngonghala+cn&cauthor_id+32360770. Accessed on 5 June 2021.

Ngonghala, C. N., Iboi, E. A., & Gumel, A. B. (n.d.-b). *Could masks curtail the post-lockdown resurgence of COVID-19 in the US?* https://pubmed.ncbi.nlm.nih.gov/?term=ngonghala+cn&cauthor_id+3281851. Accessed on 5 June 2021.

Office of the President-Kabale District; Resident District Commissioner/Chairman COVID-19 District Taskforce; 18 April 2021. Death announcements. Kabale District, Uganda.

Okware, S. I., et al. (2022). An outbreak of Ebola in Uganda. *Tropical Medicine and International Health, 7*(12), 1068–1075.

Rodriguez-Morales, A. J., MacGregor, K., Kanagarajah, S., Patel, D., & Schlagenhauf, P. (2020). Going global—Travel and the 2019 novel corona virus. *Travel Medicine and Infectious Disease, 33*, 101578. https://doi.org/10.1016/j.tmaid.2020.101578. [PMC free article] [PubMed] [CrossRef] [Google Scholar].

Schieffelin, J., et al. (2016). Clinical validation trial of a diagnostic for Ebola Zaire antigen detection: design rationale and challenges to implementation. *Clinical Trials, 13*(1), 66–72.

Segalàs, J., Tejedor, G., & Cebrián, G. (2014). (Re)Shaping knowledge: The contribution of sustainability science. In GDEE (Ed.), *The global engineer in sustainable human development*. Retrieved June 19, 2019, fromhttps://www.academia.edu/19739065/The_global_engineer_in_Sustainable_Human_Development._Re_Shaping_knowledge_the_contribution_of_Sustainability_Science

Ssejjoba, E. (2020, December 16). COVID-19: Myths, causing minimal compliance among refugees-study; some preventive measures were found not feasible. *New Vision*, p. 14.

Ssematimba, A., Nakakawa, J. N., Ssebuliba, J., & Mugisha, J. Y. T. (n.d.). *Mathematical model for COVID-19 management in crowded settlements and high-activity areas*. https://pubmed.ncbi.nlm.nih.gov/?term+mugisha+jyt&cauthor_id+33747709. Accessed on 5 June 2021.

Sturm, A., & Prudden, G. (2019). *Identifying and analysing an ethical issue*. Retrieved May 27, 2019, from https://www.physio%2D%2D-pedia.com/Identifying_and_Analysing_an_Ethical_Issue

Talks at Columbia (Producer). (2016, April 27). *Why bioethics matters* | Robert Klitzman, M.D. | Talks@Columbia. [Video file]. Retrieved June 30, 2019, from https://www.youtube.com/watch?v=6pX1gtfCDck

Tham, J. S. (2007). *The secularization of bioethics: A critical history*. Ateneo Pontificio Regina Apostolorum. https://www.academia.edu/8008282/The_Secularization_of_Bioethics_A_Critical_History

Uganda Bureau of Statistics. (2020). Statistical abstract.

Universal Declaration of Human Rights, 10 December 1948.

Universal Declaration on Bioethics and Human Rights, 19 October 2005. https://portal.unesco.org. Accessed on 30 May 2021.

Wamala, J. F., et al. (2010). Ebola haemorrhagic fever associated with novel virus strain, Uganda, 2007–2008. *Emerging Infectious Diseases, 16*(7), 1087.

Zasukhina, V. (2014, July). *Bioethics in transdisciplinary and interdisciplinary context of post-nonclassical science*. Retrieved June 19, 2019. https://doi.org/10.12955/cbup.v2.464

# Chapter 2
# Bioethics of Access to COVID-19 Vaccines

**Walter Jaoko**

**Abstract** At the height of the COVID-19 pandemic, it was evident that ending the scourge would not be feasible without the use of an efficacious preventive vaccine, resulting in an accelerated research and development of the vaccines given the seriousness of the disease. This resulted in at least seven COVID-19 vaccines being given globally, under emergency use authorization. Due to the large demand for COVID-19 vaccines, the pharmaceutical industry was not able to meet the global need for these vaccines. This resulted in distribution of these vaccines largely being based on the ability of each country to buy them. Thus, most COVID-19 vaccines were given in resource-rich countries *vis-a-*viz resource-limited ones. Was this ethical? Furthermore, even countries that could afford the vaccines faced shortages since the pharmaceutical companies were unable to manufacture them in sufficient quantities due to huge global demand. This necessitated rationing of the vaccines such that they were being given in a phased-out manner, only to certain groups first before expanding it to others. What was the ethical basis for this? Who should have been given priority over others in getting these vaccines, and why? Are there sound ethical principles that underpinned tiered access to the COVID-19 vaccines? In this chapter, the ethical basis for COVID-19 vaccine distribution both at the global and at the local ecosystem is examined.

**Keywords** Beneficence · Nonmaleficence · Priority · COVID-19 vaccines · Distribution · Access · Equity · Justice · Bioethics

W. Jaoko (✉)
KAVI-Institute of Clinical Research, University of Nairobi, Nairobi, Kenya
e-mail: wjaoko@kaviuon.org

 
L. Omutoko, W. Jaoko (eds.), *Bioethics from the Global South*, Advancing Global Bioethics 21, https://doi.org/10.1007/978-3-031-77669-4_2

# 1 Introduction

The world went through a pandemic like no other with COVID-19, which started in Wuhan, China, in December 2019 before ravaging the world. No country was spared. By the time the end of the pandemic was declared by the World Health Organization, about seven million COVID-19 related deaths had occurred globally (WHO, 2023). At the height of the pandemic, an effective, safe, accessible, and affordable preventive COVID-19 vaccine was touted as the best way of ending it, resulting in every effort being put into ensuring fast tracking in research and development of such a vaccine. The efforts included raising of research funds that is often a bottleneck in research and development (R&D) of vaccines, getting priority with regard to ethical review of research proposals, and provision of emergency use authorization (EUA) by the USA Food and Drugs Administration (FDA). EUA is a mechanism that facilitates the use of unapproved medicinal products such as vaccines during public health emergencies to prevent serious or life-threatening diseases upon certain statutory criteria being met, in the absence of adequate, approved, and available alternatives (FDA, 2020). The net effect of efforts to fast-track R&D of vaccines was that at least seven COVID-19 vaccines were developed within a record time of 9 months to the EUA level.

COVID-19 vaccine administration was unlike other vaccines in several ways. First, this was the first time that a vaccine was being given globally under EUA. Second, this was the first time that a vaccine was given in billions of doses over a very short period of time. Third, this was the first time that the world was grappling with global shortage of a vaccine and thus needing it to be distributed rationally. As a result, there were multiple bioethical issues associated with the distribution of COVID-19 vaccines to those who needed it (Wu et al., 2020), even as the world made sound public health decisions in their rollout (Priyanka et al., 2021). This chapter provides a critical review of one of these ethical issues, namely access to the COVID-19 vaccines.

# 2 Objectives

## *2.1 Research Topic*

The research topic was premised on examining if ethical principles were used in ensuring that those who need COVID-19 vaccines had access to them at the global level, and if after they had been distributed at that level, ethical principles were used in ensuring that those who needed the vaccines at the country level had access to them.

### 2.2 Research Questions

This chapter aimed to examine the following two questions, namely:

1. Were COVID-19 vaccines distributed globally in an ethical manner?
2. Did governments provide access to COVID-19 vaccines in an ethical manner?

### 2.3 Hypothesis

Ethical theories and principles were not used in determining global distribution of COVID-19 vaccines and local access by those who needed them.

## 3 Methodology

The research question for this chapter was: "What are the ethical issues that should have been taken into consideration in the distribution of COVID-19 vaccines globally and locally?" For purposes of clarity, the research question was further divided into two parts: (1) Were COVID-19 vaccines distributed globally in an ethical manner and (2) were they given at the country level in an ethical manner?

### 3.1 Inclusion

This chapter included searches of peer-reviewed articles published in the English language that either covered COVID-19 vaccines distribution at the global level or their access at a country level. The period covered was up to June 30, 2021.

### 3.2 Data Sources and Searches

Searches were performed to identify studies published on COVID-19 vaccine distribution and access, at a global or national level. The first search was a systematic review in PubMed, Embase, and Google Scholar using **keywords**: (COVID-19 vaccine) AND (distribution OR access) AND (*ethics). Search strategy included MeSH terms and free-text word variations for each database. The second search was by Google Scholar using iterating combinations of keywords such as "COVID-19," "vaccine," "distribution," "rollout," "ethics," and "bioethics."

### 3.3 Study Selection

The peer-reviewed articles were selected based on whether they addressed the issue of COVID-19 vaccine distribution at a global or local level. The first selection involved reading the article title to determine if it addressed the issue of concern. The second selection involved reading abstracts of the articles to see if they addressed the issue of COVID-19 vaccine distribution and access. References in these articles were then examined for additional articles that may have been missed, forming a three-stage selection process of identifying appropriate articles.

### 3.4 Results

A total of 415 articles were displayed in PubMed database following the search. After going through the titles of the identified articles, only 26 articles were deemed to be appropriate for the purposes of this chapter. Additional 32 articles were identified using other databases, namely Embase and Google Scholar. Of these, 13 articles appeared relevant after reading their titles. Therefore, in total, 39 articles had titles that were deemed to be appropriate for a second round of scrutiny. This round involved reading of their abstracts to determine if the article was indeed suitable for this chapter. Out of the 39 articles, 31 were relevant. These articles were further classified into two, namely those dealing with distribution of COVID-19 vaccines at the global level and those dealing with its access at the country/local level. Of these, 18 addressed issues around access to COVID-19 vaccines at the local/country level, whereas 31 articles addressed issues around the vaccines' distribution at the global level.

## 4 Description of the Facts Around COVID-19 Vaccine Distribution and Access

### 4.1 The Problem

The primary focus on the distribution of COVID-19 vaccines was premised on each nation providing them to its people, particularly to those who needed them most in the shortest time possible. Distribution of the vaccines to other parts of the world was secondary even if people there needed them more urgently. No ethical consideration was given in the distribution of these vaccines.

### 4.2 Gathered Facts

Review of published literature indicated that the focus on COVID-19 vaccines distribution by countries was primarily on getting them to their citizens, ethics notwithstanding. Ethicists had raised concern regarding this and had even suggested ways that could be used to ensure that the vaccines distribution was done in an ethical manner.

### 4.3 Clarifying Values

The values needed to ensure that COVID-19 vaccines that were distributed in an ethical manner include the following: equity, reciprocity, solidarity, and fidelity.

***Equity*** Article 10 of the Universal Declaration on Bioethics and Human Rights on equality, justice, and equity states that "the fundamental equality of all human beings in dignity and rights is to be respected so that they are treated justly and equitably" (UDBHR, 2005). In this regard, it was expected that COVID-19 vaccines would be distributed justly and equitably.

***Reciprocity*** This is one of the four principles proposed for the justification of public health interventions (Upshur, 2002), and one of the ten substantive values put forward by University of Toronto's Joint Centre Bioethics for consideration when making ethical decisions in response to a pandemic (Upshur et al., 2005). This is quite appropriate in situations such as providing access to vaccines to curb the COVID-19 pandemic.

***Solidarity*** This is described as shared practices reflecting a collective commitment to carry financial, social, emotional, or other costs to assist others (Prainsack & Buyx, 2012). It has been held as a value that supports and structures the way human beings ought to see other kinds of moral considerations and should not just simply be added to any list of values since they are not on the same level (Dawson & Jennings, 2012). It is a relational concept that allows humans to see that their condition is inextricably related to the condition of other human beings (Dawson & Jennings, 2012). This was quite evident with COVID-19 whereby controlling the pandemic in one part of the world and not the other did not make sense. Article 13 of the Universal Declaration on Bioethics and Human Rights deals with solidarity and cooperation. It states that solidarity among human beings and international cooperation toward that end are to be encouraged. This is particularly crucial in times of pandemics where global efforts are required for successful interventions such as the use of COVID-19 vaccines.

## 5 Identified Ethical Principles Applicable in the Distribution of COVID-19 Vaccines

Ethical principles that ought to have been upheld in distributing COVID-19 vaccines include:

***Autonomy*** This principle requires individuals to be allowed to choose what is best for them without coercion or undue influence. It recognizes individuals' inherent dignity independent of any other considerations such as race, socioeconomic status, gender, and health status. Based on this principle, no individual should have been discriminated against and not be prioritized to get COVID-19 vaccines for any of the considerations. Instead, individuals should have been allowed to choose to or not to be vaccinated, and that choice ought to have been respected no matter how much it was believed that the individuals needed the vaccine. This is in consort with Article 3 of the Universal Declaration on Bioethics and Human Rights which states that "human dignity, human rights and fundamental freedoms are to be fully respected" and that "the interests and welfare of the individual should have priority over the sole interest of science or society." Article 5 of the Declaration states that "the autonomy of persons to make decisions, while taking responsibility for those decisions and respecting the autonomy of others, is to be respected" (UDBHR, 2005).

***Beneficence*** The principle of beneficence emphasizes the fact that human beings should benefit maximally from any interactions such as getting COVID-19 vaccine. It is backed by Article 4 of the Universal Declaration on Bioethics and Human Rights, which states that "in applying medical practice and associated technologies, direct and indirect benefits to patients should be maximized and any possible harm to such individuals should be minimized" (UDBHR, 2005).

***Nonmaleficence*** This is the principle of "first, do no harm." It requires that individuals should not be harmed either unintentionally or deliberately. Harm in the context of COVID-19 vaccines includes denying certain individuals access to the vaccines either by making them unavailable or delaying access to them by not being categorized in the priority groups.

***Justice*** The principle of justice requires that individuals be treated equally without any form of discrimination. This principle recognizes that there will be individuals or groups who will be prioritized to get the vaccine first, but this should be done in a just manner.

***Fidelity*** This principle requires loyalty and keeping promises and commitments made to others. In the context of COVID-19 vaccines, there were commitments made by governments to their people to provide the vaccines, which should have been honored. Also, it can be assumed that governments must be committed to addressing the problems of its citizens, and this extends to ensuring that they are protected against the effects of pandemics such as COVID-19.

# 6 Clarifying Legal Rules Around the Distribution of COVID-19 Vaccines

A major challenge to global distribution of COVID-19 vaccines was the fact that they were in short supply, and pharmaceutical companies making them were unable to meet the demand for their production. Besides, since the vaccines were all newly developed, they had a patency period that needed to be observed during which nobody could make them. This was a major ethical concern. Since COVID-19 was an emergency demanding that people get immediate access to the vaccines, were the companies being ethical by denying people this access because of patency restrictions? Were the companies putting their desire for profit before duty to save lives?

## *6.1 Options to Addressing Legal Obstacles in COVID-19 Vaccine Distribution and Access*

One way that had been proposed to address this legal hurdle was to remove patency protection and allow other pharmaceutical companies to make the COVID-19 vaccine. But was this fair to the pharmaceutical companies that had invested heavily in developing the vaccines? Was there a win-win solution that would have also ensured that ethical principles were complied with?

## *6.2 Problematization of the COVID-19 Vaccine Distribution*

At the peak of the COVID-19 pandemic, each country's priority was to get the COVID-19 vaccines to as many of its citizens and as fast as possible. The performance of how a country was doing in this was assessed by the proportion of its population who had been vaccinated. This measure did not consider whether in the distribution of these vaccines, ethical principles had been used.

Problematization of the COVID-19 vaccine distribution at the global level and their access at the national level introduced an aspect of bioethics that asks questions beyond the proportion of the population reached. This includes whether distribution of the vaccines was being done equitably, fairly, and justly, and whether individuals' autonomy was being respected, and no harm was being caused.

# 7 Discussion

## *7.1 Access to COVID-19 Vaccines at Country Level*

At the height of the COVID-19 pandemic, the global demand for the vaccines could not be met by the available supplies simply because the vaccines were needed by all countries at the same time. Never in the history of mankind had there been such a demand for a vaccine. It is for this reason that countries adopted a structured phased way of giving these vaccines. Furthermore, frameworks prioritizing vaccine allocation guided by ethical considerations and informed by local risk assessment had been proposed (Fielding et al., 2021; Giubilini et al., 2020), detailing who should be given the vaccines first, who should be next, and who should be asked to wait (Gupta & Morain, 2020; Kates et al., 2021; Yang et al., 2021; Herlitz et al., 2021). But what was the criteria based on? Was there any ethical basis for following it? Was it even ethical in the first place?

Access to COVID-19 vaccines can be examined both at a local and a global level and should have involved a multidisciplinary composed of vaccine experts, policymakers, legal experts, religious leaders, ethicists, human rights activists, mathematical modelers, and medical specialists. At the local level, there were generally two main considerations in determining the pecking order in COVID-19 vaccines distribution, namely who was at the greatest risk to be infected and who was at greatest risk of getting severe disease and dying if they became infected. These considerations, largely advocated by policymakers, are based on a utilitarian bioethical theory aimed at saving the largest number of lives possible (Williams et al., 2021). And since this includes the vulnerable populations such as minority groups who are at a higher risk of being hospitalized and dying from the disease, it had the support of human rights activists and bioethicists (Abramowitz et al., 2020; Moodley et al., 2020) and is in consort with Universal Declaration of Bioethics and Human Rights Article 8, which states that "in applying and advancing scientific knowledge, medical practice and associated technologies, human vulnerability should be taken into account. Individuals and groups of special vulnerability should be protected, and the personal integrity of such individuals respected" (UDBHR, 2005).

Frontline workers were in the priority list for COVID-19 vaccination in most countries. Others had suggested that if rationing vaccines within this subgroup became obligatory for reasons such as inadequate supply of the vaccines, healthcare workers at highest risk and high-risk frontline workers in essential services should be prioritized (Moodley & Rossouw, 2021; Wu et al., 2020). There were several ethical bases for this. First, there was some evidence derived by mathematical modelers suggesting that vaccinating healthcare workers averted the greatest number of new COVID-19 infections (Ferranna et al., 2021; Young, 2020). The ethical theory underpinning this argument was utilitarianism, aiming to achieve the greatest good. Second, prioritizing healthcare workers was based on deontological ethical theory. The argument here being that since healthcare workers were highly exposed to getting COVID-19 disease by virtue of their profession while taking care of members of the society, the society had a duty to provide them with some level of protection for the risk they were taking on its behalf (Ault, 2020). The third ethical basis for prioritizing healthcare workers, also based on utilitarian theory, was that these workers were desperately needed by the society to fight COVID-19; hence, the society could not afford to lose them. For that reason, healthcare workers had to be among the first to get COVID-19 vaccines to prevent having a shortage of the workforce needed to effectively combat the pandemic. This in turn would avert more deaths in the society.

Elderly people were also prioritized to get COVID-19 vaccines. Mathematical modelers had shown that giving the vaccines to people aged 60 years and above, across countries, averted the most deaths (Bubar et al., 2021; Foy et al., 2021). Indeed, some countries had prioritized the elderly above frontline workers. For example, in its rollout, Israel started with those aged 60 year and over before healthcare workers (Rosen et al., 2021; Cylus et al., 2021). The ethical basis for prioritizing elderly people for COVID-19 vaccination was largely utilitarianism, aimed at saving the largest numbers of lives, and the number of life years saved (Craxi et al., 2021). However, others have argued that since young people contributed the greatest risk of transmitting the virus by virtue of their likelihood of being asymptomatic or only having a mild disease, they should have been prioritized to get COVID-19 vaccines as a way of cutting down transmission. In other words, vaccines should have been given to those that were more likely to transmit the infection to others (Tingley, 2020). Furthermore, although children were less likely than the elderly to get serious COVID-19 disease and were often asymptomatic, asymptomatic and presymptomatic people with COVID-19 were highly contagious and may have been responsible for a silent infection transmission (Tingley, 2020). Based on these facts, from a utilitarian perspective, there was an ethical argument for prioritizing children and young adults over elderly people to get vaccines to save the most lives, even though they were least likely to get severe COVID-19 disease. Besides, ethicists have debated whether minimizing death should have been the main determinant in prioritizing COVID-19 vaccine or society should have considered other markers such as minimization of hospitalization, thereby reducing strain on the healthcare system to cope with more severe COVID-19 diseased patients (Tingley, 2020).

In many countries, the second tire of those prioritized to get COVID-19 vaccines included adults with comorbidities. The comorbidities prioritized varied from country to country. The USA Federal government's list included conditions such as cancer; chronic kidney disease; chronic obstructive pulmonary disease (COPD); heart conditions, such as heart failure, coronary artery disease, or cardiomyopathies; immunocompromised state from solid organ transplant; obesity; sickle cell disease; smoking; type 2 diabetes mellitus; and pregnancy, per Advisory Committee on Immunization Practices (ACIP) (CDC, 2020). However, despite the Federal guidance, there was significant variation in different states on medical conditions that qualified for COVID-19 vaccine prioritization (Jain et al., 2021). Other countries such as Kenya included HIV and AIDS on the list. These conditions put the individuals at a greater risk of getting severe COVID-19 disease if they became infected. The ethical principle applied here falls within the domain of beneficence in principlism theory of bioethics since these are the ones who would benefit maximally from vaccination. Besides this, the principle of nonmaleficence comes to play since not vaccinating this vulnerable population put them at greater risk of being harmed than the public.

However, there is a debate as to which diseases should have qualified as comorbidities to be prioritized. For example, to the best of my knowledge, no country listed those with mental illness as one of the priority groups in getting COVID-19 vaccines. Yet, people with severe mental illness are known to be at an increased risk not only of COVID-19 infection but also of its related morbidity and death (Mazereel et al., 2021) and should thus have been prioritized in vaccine allocation strategies. Why were they left out? Wasn't it discrimination? Wasn't it an infringement on the rights of such people? Wasn't it unethical? Furthermore, prioritizing smokers and the obese ahead of other groups for vaccination was opposed by some people who argued that these are lifestyle choices and thus individuals with them should not be rewarded for their bad choices. Since smoking is a voluntary habit, it raised a lot of debates about fairness in prioritizing them (Read, 2021). Thus, the issue of comorbidities raised other secondary issues such as whether the comorbidity was behavior related or not. In general, it was more ethically defensible to prioritize vaccinating those with nonbehavior-related comorbidities than vice versa.

Interesting to note is that in most countries, although it was known that social determinants such as race and socioeconomic status were risk factors for COVID-19 acquisition and severity (Iacobucci, 2020), most guidance on vaccine prioritization did not explicitly consider social vulnerability at all. This defied all logic and was indeed unethical. There was no justification at all for not indicating this clearly in the guidelines for prioritization of COVID-19 vaccines while doing it for other populations that were deemed to be at similar risk of severe disease.

## 7.2 Distribution of COVID-19 Vaccines Globally

An important way of addressing the COVID-19 pandemic was for each country to achieve herd immunity against the virus. This immunity, also known as population immunity, is an indirect protection from an infectious disease realized when a population becomes immune either via vaccination or by acquiring immunity through previous infection. Achieving it with safe and effective vaccines saves lives. However, the proportion of people who must be vaccinated against COVID-19 to realize herd immunity was not known and remains an area of research. It is envisaged that this would vary with communities, populations prioritized for vaccination, and intensity of disease transmission, among other factors (WHO, 2020). To ensure that countries achieve herd immunity to COVID-19, partnerships that facilitate successful delivery of COVID-19 vaccines globally were crucial, and commitment for fair distribution of the vaccines among all countries of the world was critical (Sadeghi et al., 2021). Thus, nationalistic approaches to immunization were a major stumbling block in this regard (Forman et al., 2021). Several ethical frameworks for global distribution of COVID-19 vaccines have been proposed (Liu et al., 2020; McClung et al., 2020; Jecker et al., 2021; Sekalala et al., 2021). These frameworks emphasize principles of equity, fairness, solidarity, human rights, and inclusivity. Others have called for inclusion of reciprocity and relational equity in global response to the COVID-19 pandemic in general (Johnson, 2020; Nichol & Mermin-Bunnell, 2021).

One key determinant in the distribution of COVID-19 vaccines globally, from a public health perspective, is the transmission rate of SARS-CoV-2, the virus causing it. Mathematical models have indicated that the greatest benefits, in terms of transmission reduction, are achieved by vaccinating more people in high transmission areas than those in low infection transmission areas (Bubar et al., 2021). Since the greatest effects of the pandemic were being seen in resource-rich countries, it was logical that these are the countries that should have had access to larger quantities of COVID-19 vaccines. Indeed, this was the case. However, scrutiny of the preordered vaccines showed a worrying trend where resource-rich countries preordered much more than what their populations needed, with most of the doses being reserved for Canada, United States, United Kingdom, Australia, the European Union, and Japan. Furthermore, since local distribution of the vaccines once they arrived in different countries was influenced largely by infrastructural support, people living in resource-rich countries had access to the vaccines much earlier than those living in resource-constrained countries. The effect of this was that those living in resource-rich countries were being vaccinated and getting protection from the consequences of severe COVID-19 disease, including death, as those in poor parts of the world continued to be exposed to the risk of severe infection and death from the disease (Cioffi & Cioffi, 2021). The immense global disparities in COVID-19 vaccine access were even described in some quarters as being akin to "vaccine apartheid" (Byanyima, 2021). These concerns resulted in the setting up of a global collaboration for equitable distribution of COVID-19 vaccines, COVAX, consisting of resource-rich countries, pharmaceutical companies, foundations, and the World Bank, among others, under the direction of the World Health Organization (WHO), Global Alliance for Vaccines and Immunization (GAVI), and Coalition for Epidemic Preparedness Innovations (CEPI). The aim of COVAX was to obtain two billion COVID-19 vaccine doses before the end of 2021 and upon request ensure delivery to resource-limited countries to cover between 10 and 50% of their populations. To ensure equity, vaccine doses for each country would be allocated proportional to their populations, and no country would receive doses that exceeded 20% of its population until all COVAX beneficiary countries had reached that percentage (Dal-Réa & Camps, 2021). COVAX adopted the WHO's "fair allocation mechanism" that is based on the principle of equal proportional share per country up to when each country gets 20% of its population vaccinated, after which allocation becomes based on health needs, a framework meant to address concerns about international fairness and to check against vaccine nationalism where resource-rich nations would hoard the vaccines at the detriment of resource-limited countries (Emanuel et al., 2021; Sharma et al., 2021). COVAX facility attempted to address previous failures in responding to the COVID-19 pandemic that extended inequity, especially in resource-limited settings. The facility was in tandem with a call for global health governance of pandemic preparedness and response to shift from approaches that prioritize interests of resource-rich countries rather than enforcing equity as a central tenet in guiding the global pandemic preparedness and response strategies (Eyawo & Viens, 2020).

The ethical issue of access to COVID-19 vaccine at a global level mainly touched on equity. Was it fair for most vaccines to be distributed in resource-rich countries compared to the resource-limited countries? Should wealth have been the only consideration in accessing COVID-19 vaccines? Even before these vaccines became available through emergency use authorization, several resource-rich countries placed advance purchase orders for the vaccines undergoing clinical trials if they were found to be efficacious. As a result, although resource-rich countries constitute only 31% of the global population, they owned 51% of the world's COVID-19 vaccines. This compounded the problems of access to the vaccines beyond the inability to afford them. Even if resource-limited countries were to source funds to buy them, they would still not get them since the vaccines were unavailable for purchase. Attempts to address affordability were made through COVAX initiative that committed to ensuring that low- and middle-income countries would access COVID-19 vaccine for 20% of their populations by end of 2021 by raising funds for this purpose (GAVI, 2020; WHO, 2021). Although it was anticipated that the low- and middle-income countries would purchase the remaining portion of the required vaccines, it was unlikely that the countries would afford the vaccines at their market prices. One possibility of addressing this problem was getting generic vaccines at much cheaper prices. However, the main hurdle to achieving this was the patency laws that prohibited it, unless there was a waiver of the patency rights by vaccines' manufacturers. Although more than 100 resource-limited countries had endorsed a proposal for this matter to be discussed at the World Trade Organization (WTO), the proposal was unfortunately blocked by many resource-rich countries. A waiver would have overridden monopolies held by pharmaceutical companies from these countries but provided resource-limited countries with access to the desperately needed vaccines (UNAIDS, 2021). On the other hand, some argued that waiving patents was not the only solution to increasing access to COVID-19 vaccines by resource-limited countries. Instead, they urged manufacturers to scale up production by building more factories and earmark several billion vaccines for resource-limited countries with no access to them (Lacobucci, 2021). Whatever course of action that was to be taken, every effort should have been made to increase the manufacturing capacity of COVID-19 vaccines. But it was not just a matter of manufacturing more vaccines but of making them available to countries with very limited access to them.

A key ethical consideration in global distribution of COVID-19 vaccines is solidarity. Human beings in resource-limited countries are humans too and have dignity simply by virtue of their humanity. Death of any human being should be of equal concern, irrespective of whether it is in resource-limited or in resource-rich country. Thus, human beings were required to stand in solidarity regardless of race, sex, socioeconomic status, etc. Furthermore, it was foolhardy for those in resource-rich countries to believe that they would be safe if COVID-19 was controlled in their countries but continued to spread and to cause havoc in resource-limited countries. If COVID-19 had not been contained in other parts of the world, resource-rich countries would never have been at ease since the disease would always be lagging nearby ready to cause new epidemics in these countries.

## 7.3 Consequences of Not Receiving COVID-19 Vaccination

One concern around access to COVID-19 vaccine was with regard to the likelihood of several countries making vaccination a requirement for entry into their countries, the so-called vaccine passports. Proponents of requiring COVID-19 vaccinations for international travel argued in support of vaccine passports from the public health approach. They pointed to the fact that requirement for vaccination in international travel was not new, and many countries have for years required evidence of yellow fever vaccine or polio vaccine for entry into their lands. The ethical concern around the use of digital passports was the fear of discrimination between the vaccinated and unvaccinated. Why would people be punished for not having been vaccinated yet the vaccine was in short supply, a factor beyond their control. At that time, it was argued that if enforced, only the elderly, those in certain professions such as health-care workers and those with comorbidities from resource-limited countries would be allowed to travel since they were the ones prioritized for vaccination. Was this ethical? And even if the vaccine was available, why should one be victimized because of their choice? Aren't human beings autonomous to make their own choices? Furthermore, it was opined that this requirement had a potential of creating two systems—one for people from resource-rich countries free to travel all over and the other for those from resource-limited countries with travel restrictions. Would this be fair? There was also the risk of vaccine nationalism coming to play. Would countries insist on giving entry only to those given vaccines manufactured by them? Wouldn't this be coercion? And wouldn't it force some people to get more than one type of COVID-19 vaccine just to ease their travels? Would it be safe to be given more than one type of COVID-19 vaccines? Wouldn't this be going against the ethical principle of nonmaleficence—"first do no harm"?

Linked to the issue of the requirement of COVID-19 vaccination for international travel was the European Union proposal at the time for a requirement to have a Digital Green Certificate with details on individual's vaccination status, and recent COVID-19 PCR test results, irrespective of whether it is positive or negative (Digital COVID Certificate News, 2021). The ethical issues in this regard centered mainly around autonomy, in particular privacy and confidentiality. Similar concern had been raised in other countries, for example, Israel, in enforcing the use of "green passports" allowing its holders who had been vaccinated access to restaurants and pubs while restricting access to these facilities to nonholders. Similar proposals had been made to have vaccination certificates that would allow one to go to certain places including work (Litzinger & Hanlon, 2021). Some airlines had even proposed using immunization status for their safe air travel programs ensuring that all their staff were vaccinated (Koenig, 2021). Such requirements can be deemed to be not only unethical from the aspect of interfering with individuals' autonomy but are also an infringement on their human rights to privacy under the United Nations Declaration on Human Rights and Bioethics (UDBHR, 2005).

One way of addressing the issue of fair and ethical distribution of COVID-19 vaccines taking care of concerns highlighted above would have been to establish an ethical framework, a priori. This would include mechanisms that ensure fair and efficient allocation of the vaccines at the global level with emphasis on helping people with greatest needs, and that prioritizes giving vaccines on ethical principles of maximizing societal benefits such as averting largest number of infections and deaths from COVID-19 disease, without discriminating against the most vulnerable.

## 8 Conclusion

COVID-19 vaccines were a welcome development in the fight against the global pandemic. The short supply of the vaccines resulted in uneven distribution of the vaccines globally and to their being given in a phased-out manner. This raised ethical concerns not only of equity for resource-constrained countries but also required scrutiny of ethical basis for prioritizing some groups over others in getting the vaccines. One alternative to prioritization of COVID-19 vaccines would have been to provide them on a first-come, first-serve basis. But would this have been more ethical? It can be argued that it would not have been. Indeed, such a criterion would have raised even more ethical concerns such as not protecting the most vulnerable, which should be the focus of any ethical consideration. In addition to the ethical concerns of equity, justice, and reciprocity discussed in this chapter, distribution of COVID-19 vaccines also raised concerns about repercussions their restricted distribution may have had in the society such as discrimination caused by being denied access to pubs and restaurants and to international travel simply for not being vaccinated. These issues required the establishment of robust ethical frameworks a priori that is not just based on science but that consider ethical principles aimed at ensuring that the most vulnerable in the society is also protected.

## References

Abramowitz, P., Cobaugh, D. J., & Thompson, K. K. (2020). Transparent, equitable, safe, and effective use of COVID-19 vaccines: A societal imperative. *American Journal of Health-System Pharmacy*. https://doi.org/10.1093/ajhp/zxaa335

Ault, A. (2020). *COVID-19 vaccine should go to HCWs first*, NAS Panel Says. Medscape, October 2, 2020, https://www.medscape.com/viewarticle/938514; Young, "CDC Panel Takes on COVID Vaccine Rollout, Risks, and Side Effects."

Bubar, K. M., Reinholt, K., Kissler, S. M., Lipsitch, M., Cobey, S., Grad, Y. H., & Larremore, D. B. (2021). Model-informed COVID-19 vaccine prioritization strategies by age and serostatus. *Science, 371*(6532), 916–921. https://doi.org/10.1126/science.abe6959. PMCID: PMC7963218. PMID: 33479118.

Byanyima, W. (2021). *A global vaccine apartheid is unfolding. People's lives must come before profit*. UNAIDS. https://www.unaids.org/en/20210203_oped_guardian

CDC. (2020). *The Advisory Committee on Immunization Practices' Updated interim recommendation for allocation of COVID-19 vaccine—USA*. https://www.cdc.gov/mmwr/volumes/69/wr/mm695152e2.htm?s_cid=mm695152e2_w. Accessed on 19 May 2021.

Cioffi, A., & Cioffi, F. (2021). COVID-19 vaccine: Risk of inequality and failure of public health strategies. *Ethics, Medicine, and Public Health*. https://doi.org/10.1016/j.jemep.2021.100653. PMCID: PMC7972670. PMID: 33758775.

Craxì, L., Casuccio, A., Amodio, E., & Restivo, V. (2021). Who should get COVID-19 vaccine first? A survey to evaluate hospital workers' opinion. *Vaccines (Basel), 9*(3), 189. https://doi.org/10.3390/vaccines9030189

Cylus, J., Panteli, D., & van Ginneken, E. (2021). Who should be vaccinated first? Comparing vaccine prioritization strategies in Israel and European countries using the Covid-19 health system response monitor. *Israel Journal Health Policy Research, 10*, 16. https://doi.org/10.1186/s13584-021-00453-1

Dal-Réa, R., & Camps, V. (2021). Who should be vaccinated against COVID-19 first? *Medicina Clínica (English Edition), 156*(4), 177–179. https://doi.org/10.1016/j.medcle.2020.11.005. PMCID: PMC7825832. PMID: 33521292.

Dawson, A., & Jennings, B. (2012). The place of solidarity in public health ethics. *Public Health Reviews, 34*(4). https://doi.org/10.1007/BG03391656

Digital COVID Certificate News. (2021). Vaccine passport to travel during the COVID-19 outbreak. *COVID Passport*. https://www.covidpasscertificate.com/. Accessed 28 May 2021.

Emanuel, E. J., Luna, F., Schaefer, O., Tan, K. C., & Wolff, J. (2021). Enhancing the WHO's proposed framework for distributing COVID-19 vaccines among countries. *American Journal of Public Health, 111*(3), 371–373. https://doi.org/10.2105/AJPH.2020.306098. PMCID: PMC7893355. PMID: 33566663.

Eyawo, O., & Viens, A. M. (2020). Rethinking the central role of equity in the global governance of pandemic response. *Journal of Bioethical Inquiry, 17*(4), 549–553. https://doi.org/10.1007/s11673-020-10001

FDA. (2020). *Emergency use authorization for vaccines explained*. https://www.fda.gov/vaccines-blood-biologics/vaccines/emergency-use-authorization-vaccines-explained#. Accessed on 1 June 2020.

Ferranna, M., Cadarette, D., & Bloom, D. E. (2021). COVID-19 vaccine allocation: Modeling health outcomes and equity implications of alternative strategies. *Engineering (Beijing)*. https://doi.org/10.1016/j.eng.2021.03.014. PMCID: PMC8089031. PMID: 33968462.

Fielding, J., Sullivan, S. G., Beard, F., Macartney, K., Williams, J., Dawson, A., Gilbert, G. L., Massey, P., Crooks, K., Moss, R., McCaw, M., & McVernona, J. (2021). Constructing an ethical framework for priority allocation of pandemic vaccines. *Vaccine, 39*(5), 797–804. https://doi.org/10.1016/j.vaccine.2020.12.053

Forman, R., Shah, S., Jeurissen, P., Mark, J., & Mossialosa, E. (2021). COVID-19 vaccine challenges: What have we learned so far and what remains to be done? *Health Policy, 125*(5), 553–567.

Foy, B. H., Wahl, B., Mehta, K., Shet, A., Menone, G. I., & Britto, C. (2021). Comparing COVID-19 vaccine allocation strategies in India: A mathematical modelling study. *International Journal of Infectious Diseases, 103*, 431–438.

GAVI. (2020). *COVAX*. https://www.gavi.org/covax-facility. Accessed 28 May 2020.

Giubilini, A., Savulescu, J., & Wilkinson, D. (2020). Queue questions: Ethics of COVID-19 vaccine prioritization. *Bioethics*. https://doi.org/10.1111/bioe.12858. PMCID: PMC8013927. PMID: 33559129.

Gupta, R., & Morain, S. R. (2020). Ethical allocation of future COVID-19 vaccines. *Journal of Medical Ethics*. https://doi.org/10.1136/medethics-2020-106850. PMCID: PMC7747493. PMID: 33335070.

Herlitz, A., Lederman, Z., Miller, J., Fleurbaey, M., Venkatapuram, S., Atuire, C., Eckenwiler, L., & Hassoun, N. (2021). Just allocation of COVID-19 vaccines. *BMJ Global Health, 6*(2), e004812. https://doi.org/10.1136/bmjgh-2020-004812. PMCID: PMC7886660. PMID: 33589419.

Iacobucci, G. (2020). Covid-19: Increased risk among ethnic minorities is largely due to poverty and social disparities, review finds. *British Medical Journal, 371*. https://doi.org/10.1136/bmj.m4099

Jain, V., Schwarz, L., & Lorgelly, P. (2021). A rapid review of COVID-19 vaccine prioritization in the U.S.: Alignment between federal guidance and state practice. *International Journal of Environmental Research and Public Health, 18*(7), 3483. https://doi.org/10.3390/ijerph18073483. PMCID: PMC8036633.

Jecker, N. S., Wightman, A. G., & Diekema, D. S. (2021). Vaccine ethics: An ethical framework for global distribution of COVID-19 vaccines. *Journal of Medical Ethics*. https://doi.org/10.1136/medethics-2020-107036

Johnson, S. B. (2020). Advancing global health equity in the COVID-19 response: Beyond solidarity. *Journal of Bioethical Inquiry, 17*(4), 703–707. https://doi.org/10.1007/s11673-020-10008-9

Kates, J., Tolbert, J., & Michaud, J. (2021). *The COVID-19 "vaccination line": An update on state prioritization plans*. https://www.kff.org/coronavirus-covid-19/issue-brief/the-covid-19-vaccination-line-an-update-on-state-prioritization-plans/. Accessed on 19 June 2021.

Koenig, D. (2021, March 18). Explainer: Will you need a 'vaccine passport' to travel?" *Associated Press*. https://apnews.com/article/technology-travel-pandemics-europe-airlines-9c410bacd04c443f4ed562d1b2c3d37e

Lacobucci, G. (2021). Covid-19: How will a waiver on vaccine patents affect global supply? *British Medical Journal, 373*, n1182. https://doi.org/10.1136/bmj.n1182

Litzinger, E. N., & Hanlon, A. R. (2021). Digital vaccine passports and the legal risks for employers. *Corporate Counsel*. https://www.law.com/corpcounsel/2021/03/24/digital-vaccine-passports-and-the-legal-risks-for-employers/

Liu, Y., Salwi, S., & Drolet, B. C. (2020). Multivalue ethical framework for fair global allocation of a COVID-19 vaccine. *Journal of Medical Ethics, 46*(8), 499–501. https://doi.org/10.1136/medethics-2020-106516

Mazereel, V., Van Assche, K., Detraux, D., & De Hert, M. (2021). COVID-19 vaccination for people with severe mental illness: Why, what, and how? *Lancet Psychiatry, 8*, 444–450. https://doi.org/10.1016/S2215-0366(20)30564-2

McClung, N., Chamberland, M., Kinlaw, K., Matthew, D. B., Wallace, M., Bell, B. P., Lee, G. M., Talbot, K., Romero, J. R., Oliver, S. E., & Dooling, K. (2020). The Advisory Committee On Immunization Practices' ethical principles for allocating initial supplies of COVID-19 vaccine – United States. *Morbidity and Mortality Weekly Report, 69*(47), 1782–1786. PMCID: PMC7727606. PMID: 33237895.

Moodley, K., & Rossouw, T. (2021). What could fair allocation of an efficacious COVID-19 vaccine look like in South Africa? *The Lancet Global Health, 9*(2), e106–e107. https://doi.org/10.1016/S2214-109X(20)30474-5

Moodley, K., Rennie, S., Behets, F., Obasa, A. E., Yemesi, R., Ravez, L., Kayembe, P., et al. (2020). Allocation of scarce resources in Africa during COVID-19: Utility and justice for the bottom of the pyramid? *Developing World Bioethics, 21*, 36. https://doi.org/10.1111/dewb.12280

Nichol, A. A., & Mermin-Bunnell, K. M. (2021). The ethics of COVID-19 vaccine distribution. *Journal of Public Health Policy, 19*, 1–4. https://doi.org/10.1057/s41271-021-00291-0. PMCID: PMC8131488. PMID: 34012014.

Prainsack, B., & Buyx, A. M. (2012). Solidarity in contemporary bioethics – Towards a new approach. *Bioethics, 26*(7), 343–350. https://doi.org/10.1111/j.1467-8519.2012.01987.x

Priyanka, Choudhary, O. P., & Singh, I. (2021). Making sound public health decisions for the roll-out of COVID-19 vaccines. *Journal of Travel Medicine, 28*. https://doi.org/10.1093/jtm/taab031

Read, Z. (2021). *Why NJ's decision to vaccinate smokers has some people angry*. https://whyy.org/articles/why-njs-decision-to-vaccinate-smokers-has-some-people-angry/. Accessed on 19 June 2021.

Rosen, B., Waitzberg, R., & Israeli, A. (2021). Israel's rapid rollout of vaccinations for COVID-19. *Israel Journal Health Policy Research, 10*(1), 6. https://doi.org/10.1186/s13584-021-00440-6

Sadeghi, R., Masoudi, M. R., & Khanjani, N. (2021). The commitment for fair distribution of COVID-19 vaccine among all countries of the world. *Research in Nursing & Health, 44*(2), 266–267. https://doi.org/10.1002/nur.22112

Sekalala, S., Perehudoff, K., Parker, P., Forman, L., Rawson, B., & Smith, M. (2021). An intersectional human rights approach to prioritising access to COVID-19 vaccines. *BMJ Global Health, 6*(2), e004462. https://doi.org/10.1136/bmjgh-2020-004462. PMCID: PMC7907827. PMID: 33627362.

Sharma, S., Kawa, N., & Gomber, A. (2021). WHO's allocation framework for COVAX: Is it fair? *Journal of Medical Ethics*. https://doi.org/10.1136/medethics-2020-107152. PMCID: PMC8042584. PMID: 33837046.

Tingley, K. (2020, November 5). Who should get a Covid-19 vaccine first? *The New York Times*. https://www.nytimes.com/2020/11/05/magazine/who-gets-covid-vaccine.html

UDB and Human Rights. (2005). http://portal.unesco.org/en/ev.php-. Accessed 13 June 2021.

UNAIDS. (2021). *Rich nations vaccinating one person every second while majority of the poorest nations are yet to give a single dose*. https://www.unaids.org/en/resources/presscentre/featurestories/2021/march/20210310_covid19-vaccines. Accessed 19 June 2021.

United Nations. (2005). *United Nations Declaration on Human Rights and Bioethics*. https://en.unesco.org/themes/ethics-science-and-technology/bioethics-and-human-rights. Accessed on 29 May 2021.

Upshur, R. E. (2002). Principles for the justification of public health interventions. *Canadian Journal of Public Health, 93*, 101–103.

Upshur, R. E., Faith, K., Gibson, J. L., Thompson, A. K., Tracy, C. S., Wilson, K., & Singer, P. A. (2005). *Stand on guard for thee: Ethical considerations in preparedness planning for pandemic influenza*. University of Toronto's Joint Centre for Bioethics Pandemic Influenza Working Group.

WHO. (2020). *What is herd immunity*. https://www.who.int/news-room/q-a-detail/herd-immunity-lockdowns-and-covid-19?gclid. Accessed on 5 June 2021.

WHO. (2021). *COVAX: Working for global equitable access to COVID-19 vaccines*. WHO.int. https://www.who.int/initiatives/act-accelerator/covax. Accessed 26 March 2021.

WHO. (2023). *WHO coronavirus (COVID-19) dashboard*. https://covid19.who.int/

Williams, J., Degeling, C., McVernon, J., & Dawson, A. (2021). How should we conduct pandemic vaccination? *Vaccine, 39*(6), 994–999. https://doi.org/10.1016/j.vaccine.2020.12.059

Wu, J. H., John, S. D., & Adashi, E. Y. (2020). Allocating vaccines in a pandemic: The ethical dimension. *The American Journal of Medicine, 133*(11), 1241–1242. https://doi.org/10.1016/j.amjmed.2020.06.007

Yang, J., Zheng, W., Shi, H., Yan, X., Dong, K., You, Q., Zhong, G., Gong, H., Chen, Z., Jit, M., Viboud, C., Ajelli, M., & Yu, H. (2021). Who should be prioritized for COVID-19 vaccination in China? A descriptive study. *BMC Medicine, 19*, 45. https://doi.org/10.1186/s12916-021-01923-8

Young, K. D. (2020). *CDC panel takes on COVID vaccine rollout, risks, and side effects*. https://www.medscape.com/viewarticle/940120. Accessed on 28 May 2021.

# Chapter 3
# Restricted Patient Rights and Autonomy in the Face of COVID-19

**Mercury Shitindo**

**Abstract** The coronavirus (COVID-19), an infectious disease caused by the SARS-CoV-2 virus, emerged in China in late 2019 and rapidly evolved into a global pandemic. As of June 2024, the World Health Organization (WHO) has reported over 770 million confirmed cases and more than 7 million deaths worldwide. Kenya has also faced significant impacts, recording over 340,000 confirmed cases and more than 5500 deaths, with millions of vaccine doses administered. Beyond the immediate health crisis, the pandemic has triggered widespread human, economic, and social disruptions.

The COVID-19 pandemic necessitated unprecedented public health measures, which have posed significant ethical challenges, particularly regarding the balance between individual autonomy and the public good. This chapter explores the tension between individual rights and public health mandates, such as quarantine and vaccination requirements, assessing whether the restriction of personal autonomy can be ethically justified for the greater public good. It critically examines the ethical principles of autonomy, beneficence, nonmaleficence, and justice, and how these are applied in public health emergencies. Key sections of the chapter delineate the problem of limited autonomy versus the necessity of collective health measures, and the ethical justifications for restricting personal freedoms to prevent harm and protect public health are discussed. The role of informed consent is scrutinized, particularly where traditional consent models are challenged by urgent public health needs. Ultimately, this chapter argues for a nuanced understanding of autonomy in healthcare, advocating for policies that respect individual rights while acknowledging the ethical imperative to safeguard the health of the broader community. It underscores the importance of clear communication, trust-building, and equitable public health measures to navigate the complex ethical landscape of the COVID-19 pandemic.

M. Shitindo (✉)
Africa Bioethics Network, Nairobi, Kenya

Health Sciences, University of Zaragoza, Zaragoza, Spain
e-mail: mshitindo@africabioethicsnetworkk.org



L. Omutoko, W. Jaoko (eds.), *Bioethics from the Global South*, Advancing Global Bioethics 21, https://doi.org/10.1007/978-3-031-77669-4_3

**Keywords** COVID-19 · Autonomy · Public health ethics · Bioethics · Pandemic response · Healthcare policy · Social responsibility

## 1 Introduction

Coronavirus disease (COVID-19) is an infectious disease that is caused by a virus that began in China late 2019 and has caused a global pandemic (WHO, 2020c/5/29). Kenya had the first reported positive case of COVID-19 patient on March 13, 2020. Kenya government responded through enacting policies, directives, and laws that relate to public health, fiscal policies, and social status (behavioral) expected to mitigate and control the spread of the COVID-19 virus infection (Kenya Law, 2020). These policies are expected to mitigate and control the spread of the COVID-19 virus infection, since the condition of exclusion and suspension of individual freedoms are not adequate measures to control epidemics.

There was an executive order, which gave a directive that "any individual who doesn't follow the mandates ought to be taken to a quarantine facility" (Yusuf, 2020). Interim Guidelines on Management of COVID-19 in Kenya, Infection Prevention and Control (IPC) (2020), and Case Management define quarantine and social distancing as the separation and limited movement of healthy persons presumed exposed often at home or at a designated residential facility or hospital.

It further states that this could be applied at the individual, family, or community level. This is until it's confirmed that there is an existence of a sustained community transmission by the Ministry of Health of all confirmed COVID-19 cases identified. Contacts of cases are to be isolated either at home or in assigned facilities. Once sustained community transmission is recognized, home management of mild cases should be encouraged.

These consolidated guiding principles provide recommendations for an all-inclusive prevention and case management strategies in Kenya. Notwithstanding, adhering to these limitations has been seen as challenging to most Kenyans who consider it as unjustified, needless, a violation of their rights and most importantly interfering with their capacity for autonomy.

The nationwide mass testing drive launched by the Ministry of Health in Kenya did not attract significant participation from the majority of the population. During a televised briefing from the Ministry of Health on Sunday, May 3, 2020, the chief administrative secretary for health expressed concerns about the mass testing for COVID-19 conducted during that period. The official cautioned that the mass testing exercise was deemed a failure, despite its purpose of helping the country reduce the infection rate. Since people were not willing to be tested, there was an urgent appeal for them to voluntarily go for testing. There is indication that countries which managed to flatten their infection curve quickly relied heavily on targeted testing.

## *1.1 Context Background*

Worldwide, individuals have different reasons for resisting tests. Apart from being infected with the virus, many feared being locked up in a Kenyan quarantine center. Many equated being taken to a quarantine facility to a prison cell; likewise, the individuals who were found contradicting the laws were quarantined. If one was suspected to be COVID-19 positive, the person and their family were taken to the government-assigned facilities that, for some, were considered as a less than honorable place to live in. People were expected to spend at least 14 days once taken to the quarantine facility. Nonetheless, this period could be doubled for individuals in the centers when a person showed symptoms. The problem was that they were expected to foot the bill for their stay at the government facilities. In addition, some of the individuals complained that it was impossible to keep social distance in some of the facilities due to overcrowding, while others expressed psychological and mental anguish after the government extended their stay, one individual complained that "It's like you are condemned… it's like you are at the mercy of the government" (Mutahi, 2020).

From the abovementioned, it can be claimed that autonomy was compromised. There were complaints that chance was not given to make decisions concerning them amid enforcement of the containment measures. The ethical principle of autonomy, which has many definitions, is valued in healthcare. This chapter seeks to analyze how the aspect of autonomy is accessed during health emergencies.

## *1.2 Objectives*

1. Explaining how much control patients can have over their healthcare decisions in a public health emergency and identifying the moral considerations that should guide states during public health emergencies
2. Examining the human rights violations that are at risk in a public health emergency setting

The COVID-19 pandemic set in motion questions on how people respond to public health emergencies. The big question is, how risks and benefits should be assessed during emergencies for decision-making.

## *1.3 Significance*

It is hoped that the chapter will provide in-depth information and contribute to knowledge on how states and state organs ought to respond during emergencies. The organs include organizations that support humanitarian actions such as the Red Cross, UNHCR, and human rights groups.

## 1.4 Problem Statement

The COVID-19 pandemic compelled governments worldwide to implement stringent public health measures aimed at curbing the spread of the virus. In Kenya, these measures included mandatory quarantine, mass testing, social distancing, and lockdowns (Center for Policy Impact in Global Health, 2020). While these actions were necessary for protecting public health, they raised significant ethical, legal, and social concerns regarding the infringement on individual rights and autonomy (Johnstone & Turale, 2014; Grogan, 2022; Güner et al., 2020; Cheshmehzangi et al., 2023). This chapter delves into the ethical and social implications of these restrictions, underscoring the importance of a nuanced understanding of autonomy during public health emergencies (Dawson & Verweij, 2007).

Within the context of Kenya, there is a notable gap in knowledge regarding the impact of these measures on individual autonomy. While existing literature has often focused on the effectiveness of such measures, there remains a lack of comprehensive analysis regarding how they curtail personal freedom and decision-making (Grogan, 2022; Cheshmehzangi et al., 2023). This chapter seeks to bridge this gap by examining the intricate ways in which public health measures have restricted individual autonomy. Moreover, the tension between autonomy as a fundamental right and its limitation during public health emergencies requires exploration. While autonomy is paramount, emergencies often necessitate its restriction (Gostin & Wiley, 2020; Grogan, 2022). This chapter aims to dissect the gap between theoretical autonomy and its practical limitations during pandemics and evaluates the justifications and consequences of these limitations within the Kenyan context.

The conflict between individual good and public good emerges as another critical ethical dilemma during public health crises (Dawson & Verweij, 2007). Measures implemented to safeguard public health often prioritize communal well-being over individual rights, leading to tensions and perceptions of injustice (Childress et al., 2002; Grogan, 2022). Through case studies from Kenya, this chapter illuminates this conflict and offers an ethical analysis of the decisions made. Understanding how to balance these interests effectively without unduly compromising individual freedoms is imperative yet remains insufficiently addressed in existing research (Grogan, 2022; Cheshmehzangi et al., 2023).

Furthermore, perceptions of public health measures among Kenyans have been characterized by notions of injustice and unnecessary rights violations. This chapter endeavors to unravel the reasons behind these perceptions, drawing on surveys and interviews to provide a comprehensive understanding. Such insights are vital for enhancing public compliance and trust in future health crises, yet the existing research has not adequately explored the factors contributing to these negative perceptions (Güner et al., 2020).

Lastly, the pandemic underscored the need for robust ethical guidelines to navigate the intricate balance between public health imperatives and individual rights. This chapter critically reviews current guidelines and identifies gaps in addressing autonomy and patient rights during emergencies. Urgent attention is warranted to develop comprehensive ethical frameworks that guide healthcare providers in making balanced decisions that uphold both public health imperatives and individual freedoms, thereby preparing for future public health challenges (Dawson & Verweij, 2007; Grogan, 2022).

## 1.5 Approach

This chapter adopts a principlist approach, focusing on globally accepted bioethics principles such as autonomy, beneficence, nonmaleficence, and justice. Additionally, it integrates new analysis categories drawn from emerging concerns in bioethics, including vulnerability, equity, and sustainability, among others.

The chapter examines how measures such as mandatory quarantines, mass testing, and lockdowns have restricted individual autonomy. It evaluates whether these restrictions were justified and proportionate to the public health benefits achieved. It assesses whether the measures taken were in the best interest of public health (beneficence) and whether they minimized harm to individuals (nonmaleficence). Additionally, it explores whether the measures were implemented fairly and equitably, considering their impact on different populations and ensuring that vulnerable groups were not disproportionately affected.

## 1.6 Methodology

The methodology employed in developing this chapter involved a comprehensive literature review, utilizing a diverse array of sources to provide a robust analysis of the ethical and social implications of public health measures during the COVID-19 pandemic.

The **literature review** was conducted using multiple databases and sources. PubMed and Google Scholar were the primary academic databases used to source peer-reviewed articles, scholarly publications, and empirical studies related to patient autonomy, public health emergencies, and bioethics. Relevant policy documents, reports, and guidelines from institutional repositories, such as those from the United Nations Human Rights Council and the Universal Declaration on Bioethics and Human Rights, were reviewed to provide a foundation for the literature review. Additionally, the study included gray literature, which comprises nonpeer-reviewed sources such as reports, theses, and white papers, to gain additional insights into the subject matter.

Current and historical perspectives were enriched by reviewing articles from reputable newspapers, press releases, and statements issued during the pandemic. These sources provided real-time data and public sentiment regarding the measures implemented. Policy documents issued by the Ministry of Health in Kenya were also reviewed to understand the local context and the specific public health measures implemented during the pandemic. These documents provided crucial insights into governmental responses and their impact on patient autonomy. Various other sources, such as books on global bioethics and public health disasters (e.g., Ten Have (2016) "Global Bioethics: An Introduction" and Afolabi (2018) "Public Health Disasters: A Global Ethical Framework"), were consulted to ensure a comprehensive understanding of the ethical issues at play.

The **data collection** process involved a systematic search of the aforementioned databases and sources to identify relevant literature on the impact of public health measures on individual autonomy during the COVID-19 pandemic. The identified literature was screened for relevance and quality, which involved assessing the abstracts and full texts to ensure they provided valuable insights into the study's objectives. Key information and findings from the selected literature were extracted and categorized based on themes such as autonomy, public good, ethical guidelines, and perceptions of public health measures.

The extracted data was **analyzed** using a bioethical framework, with a particular focus on the principlist approach, examining the ethical principles of autonomy, beneficence, nonmaleficence, and justice. Additionally, the analysis addressed emerging bioethics concerns, such as issues of vulnerability, equity, sustainability, and posttrial access, in the context of the COVID-19 pandemic.

## *1.7 Self-determination in Light of a Pandemic*

Self-determination, a cornerstone of individual autonomy, is fundamentally challenged during a pandemic such as the COVID-19 crisis (World Health Organization, 2020a, b, c). This principle emphasizes the right of individuals to make decisions about their own lives, free from external coercion (Beauchamp & Childress, 2001). However, the balance between individual rights and collective responsibility becomes particularly complex in the context of a public health crisis.

The COVID-19 pandemic has necessitated unprecedented public health measures, including lockdowns, mandatory quarantines, and vaccination campaigns (World Health Organization, 2020a, b, c). These measures, while essential for controlling the spread of the virus and protecting public health, often infringe upon individual autonomy. For instance, mandatory quarantine measures can be perceived as a violation of personal freedom, yet they are justified on the grounds of preventing harm to others (Engelbrecht, 2014a, b). This creates a tension between respecting individual self-determination and ensuring the welfare of the community.

In light of this, the concept of self-determination must be re-evaluated. It cannot be viewed in isolation but rather in the context of societal needs and the potential consequences of individual actions on public health. The ethical justification for limiting self-determination hinges on the principles of beneficence and nonmaleficence, which prioritize actions that benefit others and avoid harm (Beauchamp & Childress, 2001). During a pandemic, these principles may necessitate restrictions on personal freedoms to safeguard the health of the broader population.

Furthermore, transparency and communication are crucial in maintaining trust and compliance with public health measures. Authorities must clearly explain the rationale behind restrictions and engage with the public to address concerns and misconceptions. This approach not only respects individuals' rights to information but also fosters a sense of collective responsibility and solidarity (World Health Organization, 2020a, b, c).

In addition, the principle of justice must guide the implementation of public health measures to ensure they are equitable and do not disproportionately burden vulnerable populations. Equitable access to healthcare resources, accurate information, and support systems are vital in upholding the dignity and self-determination of all individuals, particularly those in marginalized communities (Ten Have, 2016).

## 2 Autonomy as a Fundamental Right

### *2.1 Intrinsic Value of a Person vs. Autonomy as a Fundamental Right*

The intrinsic value of every individual forms the cornerstone of bioethical considerations, juxtaposed against the fundamental right of autonomy. Within the global bioethics discourse, autonomy stands as one of the four core principles, alongside beneficence, nonmaleficence, and justice (Ten Have, 2016). It represents the right of individuals to make decisions about their own lives, including choices related to healthcare.

In contrast, the COVID-19 pandemic has brought to the forefront the tension between individual autonomy and public health imperatives. While autonomy is crucial, pandemic responses often necessitate restrictions on individual freedoms to safeguard public health. This tension underscores the need for a nuanced understanding of autonomy and its limitations during public health emergencies.

One of the key challenges lies in reconciling the intrinsic value of each person with the necessity of limiting autonomy to mitigate the spread of infectious diseases. This necessitates a careful examination of how autonomy intersects with other ethical principles, such as beneficence and justice, particularly in the context of resource allocation and public health interventions (Afolabi, 2018).

Autonomy is fundamental to the intrinsic value of a person, their worth, their centrality, and what differentiates them from other creatures. When this notion of a person is reduced through any means regardless of why has to be critically examined. In this chapter, the question of how an individual's autonomy is affected by the treatment, preventive measures adopted in the tackling of the COVID-19 pandemic, how do they affect an individual's rights will be deliberated centrally through the comprehension of what is autonomy, how its identified with human rights, and what actions result in its infringement.

In examining autonomy, this chapter delves into the philosophical understanding of autonomy, its current definition according to human rights organizations such as the United Nations Human Rights Council (UNHRC, 1948), and the Universal Declaration on Bioethics and Human Rights (UDBHR, 2005). The chapter endeavors to utilize a bioethical analysis to understand the relationship of autonomy and human rights in the context of a public health emergency such as the current COVID-19 pandemic (World Health Organization, 2020a, b, c).

## 2.2 The Bioethical Problem

The bioethical landscape is confronted with a complex dilemma: how to balance the intrinsic value of every individual against the fundamental right of autonomy, especially in the context of the COVID-19 pandemic (World Health Organization, 2020a, b, c). This problem becomes more pronounced as public health measures, including quarantine mandates, intersect with individual freedoms (Center for Disease Control and Prevention, 2020).

Coronavirus disease (COVID-19), an infectious disease caused by a virus that began in China late 2019, has led to a global pandemic (World Health Organization, 2020a, b, c). The world, including Kenya, has been adversely affected by the pandemic. There are continued efforts to produce effective vaccines or treatments for COVID-19, with many ongoing clinical trials evaluating potential treatments while simultaneously trying to reduce and control its spread (World Health Organization, 2020a, b, c).

Kenya has been at the forefront in fighting the pandemic in terms of laying out measures that will help control the spread, enacting emergency policies, and working with the Ministry of Health, healthcare providers, and the security arm in mitigating this (Ministry of Health, Kenya, 2020). However, this has also led to the feeling of a lack of autonomy for citizens in the interests of public health, with many Kenyan's considering some of the limitations as unjustified, needless, a violation of their rights, and most importantly, interfering with their capacity for autonomy (Human Rights Watch, 2020).

The specific problem revolves around the tension between preserving individual autonomy and ensuring public health during a pandemic. While autonomy is considered a fundamental right, its exercise may conflict with efforts to contain the spread of infectious diseases (Beauchamp & Childress, 2013). This raises questions about the extent to which autonomy can be restricted in the interest of public health while still upholding ethical principles such as beneficence, nonmaleficence, and justice (Beauchamp & Childress, 2013).

Moreover, the pandemic has underscored disparities in access to healthcare and resources, exacerbating existing inequities. Vulnerable populations, including the elderly and marginalized communities, may face heightened risks and limited autonomy in decision-making. This highlights the need for a nuanced approach that considers the intersecting factors of vulnerability, equity, and sustainability in bioethical decision-making (Ten Have, 2016; Afolabi, 2018).

### 2.2.1 Autonomy

Autonomy, as a cornerstone of bioethics, embodies the principle that individuals possess the right to make decisions about their own lives and bodies (Beauchamp & Childress, 2001). However, in the context of a public health crisis like the COVID-19 pandemic, the exercise of autonomy can encounter significant challenges and limitations (World Health Organization, 2020a, b, c).

The problem arises when individual autonomy clashes with the broader societal imperative to safeguard public health. Mandates such as quarantine orders, travel restrictions, and vaccination requirements may curtail individual freedoms in the interest of containing the spread of the virus and protecting vulnerable populations (Center for Disease Control and Prevention, 2020). This tension between individual autonomy and public health imperatives underscores the need for careful ethical deliberation and decision-making (Beauchamp & Childress, 2001).

From the abovementioned, it may very well be reasoned that there is a similarity to an absence of autonomy. There are a few people who may feel they are not being offered the chance to pick what befalls them and how. The ethical principle of autonomy, which has many definitions, is highly valued in personal healthcare decision-making. However, during health emergencies, the aspect autonomy could become difficult to access (Beauchamp & Childress, 2001).

The COVID-19 pandemic has brought this aspect into the limelight, questioning exactly how much control patients should have over their healthcare decisions. Bioethicists generally advocate for maximizing patient autonomy and giving individuals as much control as is achievable over their healthcare decisions (Beauchamp & Childress, 2001). In any case, when do we realize that a patient will be unable to completely control their healthcare choices? Especially with a public health emergency setting, how do we set ethical guides that can be adhered to in the face of uncertainty, confusion, panic, and misinformation as would be present in a public health emergency? Establishing ethical guidelines becomes crucial in these circumstances, considering factors like uncertainty, confusion, panic, and misinformation that are guided by principles like justice, autonomy, beneficence, and nonmaleficence (Beauchamp & Childress, 2001).

The very nature of being a physician involves an obligation of beneficence to patients. Hence, morally, physicians would have a greater obligation to help patients than nonphysicians. Beauchamp and Childress (2001) described beneficence as the principle of a person acting with the best interest of another in mind whereas nonmaleficence being the principle that "above all, do no harm," which is also detailed in the Hippocratic Oath. Accordingly, in the medical care settings we need to determine the patient's perspectives on what counts and does not count as goods to be pursued or harms to be avoided. The medical fraternity as a whole also has an implicit agreement with society to provide medical assistance in the midst of a health emergency (Beauchamp & Childress, 2001).

John Stuart Mill expressed that the idea of respect for autonomy involves the ability to think, choose, and act on the basis of such thought and decision uninhibitedly and freely. Mill's principle of autonomy, also known as the principle of liberty, advocates for personal freedom as long as it does not harm others (Mill, 1974). The principle of not causing harm to others (known as Mill's "harm principle") gives the grounds for the ethical right of a patient to decline medical treatment and for a physician to abstain from intervening against the patient's desires. Still, Mill held that it was acceptable to prevent people from harming themselves so long as their action is being done while the person is not fully informed (Mill, 1998).

Moreover, autonomy is not evenly distributed across society, with marginalized groups often facing barriers to exercising their autonomy fully. Factors such as socioeconomic status, access to healthcare, and educational attainment can influence an individual's ability to make informed choices about their health and well-being. Addressing these disparities is essential for ensuring that autonomy is upheld equitably for all members of society (Ten Have, 2016; Afolabi, 2018).

In navigating the complexities of autonomy during a public health crisis, it is crucial to strike a balance between respecting individual rights and promoting the common good. This requires a nuanced approach that takes into account the unique circumstances of each situation while upholding ethical principles such as beneficence, nonmaleficence, and justice (Beauchamp & Childress, 2001).

### 2.2.2 Human Rights

Relationship of Autonomy and Human Rights

This section investigates the interplay between autonomy and human rights during the COVID-19 pandemic. It examines how public health measures, while necessary for controlling the virus, may have infringed upon individual rights and analyzes whether these infringements were justified in the context of protecting the greater public good and how they can be mitigated in future public health emergencies.

Autonomy is an essential part of human rights. This means human beings have a right to have autonomy so that they can feel free to make decisions. In essence, this is the right to personal autonomy and to physical and psychological integrity. Human rights have progressively been advanced as an important framework for bioethics and are also considered to be a standard approach to biomedical ethics. Beauchamp and Childress developed the four principles—autonomy, nonmaleficence, beneficence, and justice (Beauchamp & Childress, 2001) in their book titled "Principles of Biomedical Ethics." These principles autonomy (self-determination), beneficence and nonmaleficence, and justice are aimed at resolving ethical issues where each needs to be evaluated and balanced in determining an ideal course of action. These principles need the initiation of a conversation about the needs and desires of a patient or, in the case of justice, members of community. They are merely intended to guide decision-making process. Once the wishes of the patient are established, then it becomes possible to protect his or her autonomy. Healthcare professionals face the ongoing task of balancing patient autonomy with other ethical considerations in their daily work. They must uphold patients' right to make informed decisions while considering the potential impact on their well-being and adhering to professional responsibilities. This complex task involves navigating the interplay between autonomy and ethical principles to provide the best possible care for patients.

### 2.2.3 Human Dignity

Autonomy, fundamental to the intrinsic value of persons, their worth, their centrality, and what differentiates them from other creatures, is a critical aspect of human dignity (Beauchamp & Childress, 2001). When this notion of personhood is reduced through any means, it affects an individual's dignity. This is one's right to a private life that will protect their dignity and autonomy, i.e., the right to be independent and make their own decisions about their life (United Nations, 1948).

The concept of autonomy is deeply rooted in human rights. The Universal Declaration of Human Rights (UDHR) emphasizes the importance of autonomy in the context of human dignity and well-being. Article 3 of the UDHR states that everyone has the right to life, liberty, and security of person. This right to life includes the right to make decisions about one's own body and health, which is a fundamental aspect of autonomy (United Nations, 1948).

The concept of autonomy is also closely tied to the principle of self-determination. This principle is enshrined in Article 1 of the UDHR, which states that all human beings are born free and equal in dignity and rights. This principle emphasizes the importance of individuals having the capacity to make decisions about their own lives and to exercise their rights without interference from others (United Nations, 1948).

In the context of healthcare, autonomy is essential for ensuring that patients have the right to make informed decisions about their own care. This includes the right to refuse treatment, to choose their own healthcare providers, and to make decisions about their own health and well-being. The principle of autonomy is also closely tied to the principle of informed consent, which requires healthcare providers to obtain the informed consent of patients before performing any medical procedures (Beauchamp & Childress, 2001).

The concept of autonomy is also influenced by various theoretical frameworks, such as the social contract theory and the concept of moral agency. According to social contract theory, individuals enter into a contract with each other to form a society and to establish rules and norms that govern their behavior. This contract is based on the principle of mutual respect and the recognition of each other's autonomy (Rousseau, 1762).

The concept of moral agency is also closely tied to the concept of autonomy. Moral agency refers to the capacity of individuals to make decisions and to act on their own behalf. This capacity is essential for ensuring that individuals have the ability to exercise their autonomy and to make decisions about their own lives (Beauchamp & Childress, 2001).

The bioethical problem within the realm of human rights is multifaceted, especially concerning the tension between individual liberties and collective well-being during a public health crisis like the COVID-19 pandemic. Human rights frameworks, rooted in principles of dignity, equality, and nondiscrimination, provide a lens through which to examine these complexities (United Nations, 1948).

At its core, the clash between individual rights and public health imperatives highlights the need for a delicate balance between respecting fundamental freedoms and safeguarding the welfare of communities. While individuals have the right to autonomy and self-determination, this must be weighed against the potential harm posed by unchecked transmission of infectious diseases (World Health Organization, 2020a, b, c).

Furthermore, human rights discourse emphasizes the importance of equity and justice in decision-making processes. Vulnerable populations, including the elderly, minorities, and socioeconomically disadvantaged groups, are often disproportionately affected by public health measures. Ensuring that their rights are upheld and their voices are heard is essential for promoting fairness and inclusivity in pandemic response efforts (United Nations, 1948).

Drawing upon theoretical frameworks such as utilitarianism, which prioritizes the greatest good for the greatest number, and the capabilities approach, which focuses on enhancing individuals' ability to lead a flourishing life, can offer insights into reconciling conflicting ethical priorities. These perspectives underscore the need to balance individual freedoms with collective responsibilities in pursuit of a just and equitable society (Mill, 1859; Nussbaum, 2000).

Navigating the intersection of human rights and public health ethics requires a nuanced understanding of ethical principles, social justice considerations, and the unique circumstances of each situation. By upholding human rights principles while promoting the common good, policymakers and healthcare professionals can strive to mitigate the bioethical challenges posed by pandemics like COVID-19 (World Health Organization, 2020a, b, c).

### 2.2.4 Social Responsibility

Looking at how the roles and responsibilities, preparedness, and response of a state, a nation, and a government come into play, it is evident that all areas of society are engaged in pandemic preparedness and response. The national government, such as Kenya's, is the natural leader in charge of the general coordination and communication efforts (Ministry of Health, Kenya, 2020). National preparedness and response to emergencies, such as pandemics or natural disasters, are a collective responsibility that involves the entire society. This approach to the COVID-19 pandemic underlines the significant roles played not only by the Ministry of Health but also by all other sectors, individuals, families, and communities, in mitigating the effects of the impacts of COVID-19 (World Health Organization, 2020a, b, c).

Seedhouse (2009) notes that even autonomous people with self-governing capacities sometimes fail to govern themselves in the making of particular decisions because of temporary constraints caused by illness or depression or because of ignorance, coercion, or other conditions that restrict their options. Hence, autonomy may in itself benefit the patient, but it may sometimes disadvantage him or her as well. For example, persons who refuse to go for quarantine when they are exhibiting symptoms of COVID-19 may not just be putting their health and those of their loved ones at risk, but it may also be difficult for health professionals to carry out their moral duties of doing goods (concept of beneficence), that is, to adequately treat any severe symptoms and avoid increased cases of fatalities (Beauchamp & Childress, 2001). Furthermore, Rogers (2002a) posits that, when provided with information and the opportunity for greater involvement in making decisions, consumers generally become warier of the treatments offered and make more conservative decisions.

Social responsibility emerges as a crucial aspect within the bioethical discourse surrounding the COVID-19 pandemic, intertwining with notions of solidarity, reciprocity, and community well-being. This dimension of the bioethical problem underscores the collective duty to prioritize actions that benefit society as a whole, particularly in times of crisis (United Nations, 1948).

Central to the concept of social responsibility is the recognition that individual actions can have significant consequences for public health outcomes. While individual autonomy is a fundamental human right, it must be balanced with the collective responsibility to protect the health and well-being of society as a whole. Ethical frameworks such as communitarianism emphasize the importance of collective values and shared responsibilities in guiding decision-making processes (Etzioni, 1993).

The concept of social responsibility is rooted in the idea that individuals and institutions have a moral obligation to contribute to the common good and to promote the welfare of others. In the context of a pandemic, this means that individuals have a responsibility to follow public health guidelines, such as wearing masks, practicing social distancing, and getting vaccinated, even if these measures may infringe on their personal freedoms to some extent (World Health Organization, 2020a, b, c).

Furthermore, social responsibility entails a commitment to equity and inclusivity, ensuring that public health measures do not exacerbate existing disparities or marginalize already vulnerable groups. Utilizing principles of distributive justice, policymakers can strive to allocate resources fairly and address the needs of underserved communities, thereby promoting greater social cohesion and solidarity (Rawls, 1971).

The COVID-19 pandemic has highlighted the interconnectedness of global health systems and the importance of international collaboration in addressing shared challenges. From a global bioethics' perspective, social responsibility extends beyond national borders, necessitating cooperation and solidarity among countries to ensure equitable access to vaccines, treatments, and essential healthcare services (World Health Organization, 2020a, b, c).

Governments also have a responsibility to protect the health and safety of their citizens during a pandemic. This includes implementing evidence-based public health measures, such as lockdowns and travel restrictions, to limit the spread of the virus. However, these measures must be proportionate, necessary, and implemented in a fair and equitable manner (United Nations, 1948).

Healthcare workers also have a social responsibility to provide care to those in need, even at personal risk. This responsibility is grounded in the ethical principles of beneficence and nonmaleficence, which require healthcare workers to do good and avoid harm (Beauchamp & Childress, 2001).

The concept of social responsibility is also influenced by various theoretical frameworks, such as utilitarianism and virtue ethics. Utilitarianism holds that the right course of action is the one that maximizes overall well-being and minimizes harm. In the context of a pandemic, this would suggest that individual freedoms may need to be limited to protect the greater good (Mill, 1863). Virtue ethics, on the other hand, focuses on the character of the moral agent rather than on the consequences of their actions. In this view, individuals have a responsibility to cultivate virtues such as compassion, courage, and civic mindedness and to act in accordance with these virtues even in challenging circumstances (Aristotle, 350 BCE).

By embracing social responsibility as a guiding principle in pandemic response efforts, we can foster solidarity, promote equity, and safeguard the well-being of communities worldwide. By recognizing our interconnectedness and collective obligations, we can navigate the complex ethical terrain posed by public health emergencies like COVID-19 more effectively (World Health Organization, 2020a, b, c).

## 3 The Kenya Perspective

### *3.1 The Situation in Kenya*

Many Kenyans seem to have varied reasons behind their reluctance in availing themselves for the COVID-19 tests. One of the reasons reverberates with the aspect of autonomy. Many do not actually comprehend what COVID-19 is and its relevance to their lives (Chau, 2021). After all, the disease cannot be seen. The people who were "claimed" to test positive were not known to everybody, therefore creating room for doubt. Others associate it with causing problems in their life socially, economically, and psychologically. Then, there was the dread of imagined pain that could be "experienced" in the testing process. This is because some of the individuals who had undergone the test before claimed it was painful. Additionally, others were afraid of contracting the virus during testing since they could be exposed to the ones who getting tested. This was also experienced at hospitals which reported low hospital visits for health services, citing fear of contracting the virus, or being assumed to be infected. The Ministry of Health went ahead and assured Kenyans that all hospitals had taken steps that would ensure that other services were offered without putting people at risk (Wasike, 2020).

From the above, it may very well seem to be an indication of a lack of autonomy. There are some individuals who may feel they are not being and assured offered the opportunity to pick what befalls them and how.

Some individuals may feel that they are not being given the assurance or opportunity to have control over the events and outcomes that affect their lives. They may perceive a lack of autonomy, where their choices and decisions are disregarded or overridden by external factors or authorities. This can lead to feelings of frustration, helplessness, and a sense of being powerless in shaping their own circumstances. Recognizing and respecting individuals' autonomy are essential in empowering them to make decisions that align with their values and preferences, promoting their overall well-being and sense of agency. Autonomy is critical in personal healthcare decision-making. While it is right to say that in certain conditions, other contemplations outweigh the desires of an individual like in instances where their decisions may cause avoidable harm to others—autonomy should be thought of more comprehensively. Autonomy could be limited in instances where exercising it causes harm to others or harms the patient and or person or if it is exercised, it will violate the healthcare team's medical conscience. Patient well-being and autonomy go hand-in-hand. Autonomy ought not to necessarily be viewed as "patient control of decision-making" but as a clinical reality which comprises of education, discussion, and concern for the patient's well-being (Hospital News, 2020).

It is helpful for healthcare professionals to consider autonomy, not as an unimportant principle or as something that is entirely lost if a right to choose is deprived but rather as a matter of the degree to which it is regarded, with the end aim of providing deferential patient care. The ramifications are that once information pertinent to treatment is made available, then the patient is considered capable of making healthcare decisions, at which point the healthcare providers recommending treatment should not stop him/her from their decision unless if by respecting this decision it will cause harm to others or seriously undermine the patient's well-being (Hospital News, 2020). Consideration of patient autonomy should consistently be the starting point when looking for ways in enhancing patients' dignity. Past this, the issues must be resolved using suitable moral reasoning, a comprehensive assessment of the situation, clear communication, respect empathy, and personal judgment (Hospital News, 2020).

The question to answer is how do we place autonomy as a principle in the current situation whereby the government mandates people to go for quarantine for "their own good"? Is this an action for public good or there is a violation of one's fundamental rights, autonomy, and beneficence? Autonomy and individual responsibility assume that the individual is capable of making autonomous decisions while taking responsibility for those decisions and in respect of the autonomy of others.

## 3.2 Autonomy as a Central Value

Autonomy is a central value in medical and health ethics. It is conceivable that autonomous individuals are often in the best position to determine what would be good/acceptable and bad/not acceptable for them (Sumner 1996). Consequently, it is arguable that there is valid justification to consider patients' autonomy to have instrumental value in their healthcare decisions. Looking at where the role of public good comes to play and at which point, we should consider to override this principle.

Balancing the principle of public good with individual autonomy is a complex ethical consideration. In certain situations, overriding individual autonomy may be necessary to promote the well-being of the larger population, especially during public health emergencies. However, such decisions should be carefully evaluated, taking into account factors like proportionality, necessity, and transparency. Efforts should be made to communicate the reasoning behind these decisions and ensure fairness and equity in their implementation. As such, the principle of autonomy does acknowledge the positive duty on a healthcare professional to respect the choices of a patient. The common notion is to always ensure the protection of a person's freedom, privacy, and integrity. Trust and communication remain the basis of the relationship between patients and healthcare providers.

Engelbrecht (2014a, b) posits that balance has to be struck between the good of an individual and the good of the community when looking at the concept of autonomy in the practice of healthcare. Often, finding this balance presents challenges to healthcare practitioners in Kenya, especially those in the public sector. In essence, autonomy is an indicator of "one's legal and mental capacity to comprehend and make an informed decision" (O'Neill 2002). Consequently, for an individual to adequately exercise personal autonomy, they need to have the capacity to understand the information available and its appropriateness. Providing all relevant information and assistance is a key ethical responsibility of healthcare representatives. Legislators play the role of providing the legal framework supporting autonomy. First, to be autonomous (in a real sense a self-lawmaker), a person must have adequate information to explore and examine all options pertinent to the healthcare decision they need to make. This can be considered specialized knowledge, which is most often beyond the scope of most patients and therefore often relies on healthcare providers to present them with the information they lack (often in a simplified version).

Moreover, albeit a patient may fully comprehend the medical treatments and consent to it, it can hardly ever be held that the consent and autonomy have truly been demonstrated. It's possible that a patient may almost never fully understand all the medical terms, procedures, treatments, and the consequences. Caplan (2014) contends that informed consent is inherently limited in scope, arguing that a patient cannot fully comprehend or appreciate all potential risks associated with medical treatments and procedures.

## 3.3 *Limitations in Personal Autonomy*

Epstein (2006) asserts that having the freedom to self-govern does not necessarily mean that a person will instantaneously have the capacity and the opportunity to exercise the right of self-rule. However, the limitation in human, medical, and technical resources does not justify abandoning the ethical and legal obligation of providing care to a patient and ease their suffering, always observing the informed consent and the patient's human rights. Medical intervention can be justified in cases where there is evidence of dangerous behavior and is done to protect the individual's well-being and the safety of others. Such an intervention is permissible in order to prevent any harm to others or to self. Society confers professional autonomy, and it is limited in comparison to the protection of a patient's autonomy and human rights. The respect for autonomy has only prima facie standing and competing moral considerations can sometimes override it. In instances when a person's decisions endanger public health and thus potentially harm others, others can justifiably curb their exercising of autonomy.

Zolkefli (2017) postulates that, to be genuinely autonomous, we are required to take seriously the social implications of our autonomous decisions. Placing too much emphasis on the promotion of individual patient autonomy, particularly when such decisions are actually made alone, carries the risk that we might forget either the interests of others or the wider public interest (Zolkefli 2017). In this case, it would greatly benefit the society if people would cooperate with the medical practitioners and by extension the state, beginning with the populace willingly accepting to undertake COVID-19 testing.

Personal autonomy, defined as the capacity to make informed and uncoerced decisions, is a foundational principle in bioethics. However, this autonomy is not absolute and can be limited under certain circumstances, particularly when individual decisions pose risks to public health or infringe on the rights of others. The COVID-19 pandemic has underscored the need to balance individual freedoms with collective responsibilities, presenting a complex bioethical challenge.

### 3.3.1 Ethical Justifications for Limiting Autonomy

One key justification for limiting personal autonomy is the principle of nonmaleficence, which obligates healthcare providers and policymakers to prevent harm. During the COVID-19 pandemic, measures such as quarantine, isolation, and mandatory vaccination programs were implemented to curb the virus's spread. These interventions, while restrictive, are ethically justifiable to prevent harm to others and protect public health (Beauchamp & Childress, 2013).

The principle of beneficence further supports the limitation of autonomy by emphasizing actions that promote the well-being of individuals and the community. Public health measures that restrict individual freedoms can be seen as acts of beneficence, aiming to reduce morbidity and mortality rates associated with the pandemic. These measures are particularly crucial when voluntary compliance is insufficient to achieve the desired public health outcomes.

### 3.3.2 Legal and Social Constraints

Legally, the limitation of personal autonomy is often grounded in the state's duty to protect public health and safety. Public health laws and emergency powers can authorize interventions that restrict individual freedoms during crises, provided these actions are necessary, proportionate, and time-limited. This legal framework ensures that autonomy is not curtailed arbitrarily and that any restrictions are justified by a legitimate public health goal (O'Neill, 2002).

From a social perspective, the concept of social responsibility implies that individuals have a duty to consider the impact of their actions on others. In a pandemic, this translates to a moral obligation to adhere to public health guidelines, such as wearing masks, practicing social distancing, and accepting vaccinations. The ethical notion of solidarity underscores this responsibility, highlighting the interconnectedness of individuals within a community and the collective effort required to manage public health threats (Zolkefli, 2017).

### 3.3.3 The Role of Informed Consent

Informed consent is a cornerstone of respecting autonomy in healthcare. However, during a pandemic, the traditional model of informed consent may face challenges due to the urgent need for widespread public health measures. While it is essential to provide individuals with information about the risks and benefits of interventions, there may be situations where consent cannot be fully obtained, such as during mandatory quarantines or isolation orders. In these cases, transparency, clear communication, and efforts to minimize the impact on personal liberties are crucial to maintaining ethical standards.

### 3.3.4 Balancing Autonomy and Public Health

The balance between personal autonomy and public health is a delicate one. Ethical decision-making in this context requires a careful assessment of the potential benefits and harms of restricting autonomy, as well as consideration of equity and justice. Vulnerable populations, who may be disproportionately affected by both the pandemic and the measures taken to control it, must be given particular attention to ensure that their rights and well-being are protected.

While personal autonomy is a fundamental right, it is not without limitations, especially in the context of public health emergencies. Ethical principles such as nonmaleficence, beneficence, and social responsibility provide a framework for justifying these limitations. Legal and social mechanisms further support the need to balance individual freedoms with the collective good, ensuring that measures taken are necessary, proportionate, and equitable.

# 4 Recommendation

The concept of autonomy in the practice of healthcare has to find a balance between the good of an individual and the good of the community (Engelbrecht, 2014a, b). This tension presents challenges to healthcare practitioners in Kenya, especially those in the public sector. In essence, autonomy is a manifestation of "one's legal and mental capacity to understand and make an informed decision" (O'Neill 2002). Therefore, to exercise personal autonomy, one needs the capacity to understand what is available and whether it is appropriate for one's purpose. Providing information and assistance is the key ethical responsibility of healthcare representatives. Providing the legal framework supporting autonomy is the role of legislators.

First, to be autonomous (literally a self-lawmaker), an individual must have adequate knowledge to explore and examine all options relevant to the healthcare decision that needs to be made. This specialized knowledge is beyond the scope of most patients, so they must rely on healthcare professionals to present them with the information in a simplified version. Furthermore, although a patient may fully understand the medical treatments and consent to it, it can seldom be said the consent and autonomy are truly manifested. A patient will almost never fully grasp all the medical procedures and consequences. In this regard, Caplan (2014) argues that consent is "inherently limited." A patient is not in the position to fully predict, let alone comprehend or appreciate all the risks associated with the medical treatments and or procedures.

## *4.1 Enhancing Patient Autonomy Through Information*

### 4.1.1 Provision of Information

Patients may possess significant knowledge about their illness, but often lack a comprehensive understanding of all relevant aspects. The COVID-19 pandemic has exacerbated this issue, with a deluge of information and misinformation making it challenging for individuals to make well-informed and rational decisions. It is the responsibility of healthcare professionals to provide accurate, relevant, and clear information to patients. This enables patients to make informed decisions about their medical treatment, ultimately enhancing their autonomy and dignity.

In this context, the Ministry of Health or a designated health institution should be responsible for disseminating information to the public. While social media and televised advertisements can be effective, they primarily reach individuals who already have access to these platforms and some understanding of the pandemic and miss others who do not have the same access. A more comprehensive approach is needed to ensure information reaches the common "mwananchi"—including everyone from the general populace, children, vulnerable groups, minorities, and individuals living in low-income and rural areas. This might involve utilizing community leaders, local radio broadcasts, posters, and direct community engagement to ensure that all segments of the population are adequately informed.

Using a person or medium knowledgeable to them would build trust that would give them confidence to make informed decisions on their health. This will also make it possible for everyone to understand the importance of social distancing, safety measures, wearing of masks, lockdown, and others—and that it will require them to bear with a difficult period economically too, albeit only for a period of time. This would definitely eliminate the problem of the unwillingness of people getting tested, or abiding with the safety measures put in place. Providing reliable information on the risk, severity, and progression of a pandemic and the effectiveness of interventions used during this pandemic requires clear communication protocols and dedicated resources.

## *4.2 Promoting Social Responsibility*

### 4.2.1 Balancing Individual Rights and Collective Responsibilities

To protect our collective right to health in the current pandemic situation, we need to balance our individual rights with collective responsibilities. In order to fully implement human rights, we need to place more emphasis on the responsibility of all actors, and not just the government and its machinery, to take action together to make sure rights are enjoyed. The health sector (including public health and both public and private healthcare services) has a natural leadership and advocacy role in such a pandemic preparedness and response efforts, advocating for a culture of solidarity, where community members understand that their actions impact the well-being of others. Educational programs can help foster this sense of collective responsibility, making it clear that individual sacrifices are necessary for the greater good. Public health campaigns should emphasize the importance of social responsibility, encouraging individuals to adhere to measures such as vaccination, mask-wearing, and social distancing.

## *4.3 Legal and Ethical Frameworks*

### 4.3.1 Strengthening Legal Protections

Legislators should ensure that public health laws are clear, equitable, and transparent. These laws must provide a framework for implementing public health measures that limit personal autonomy, such as quarantine and isolation, only when necessary and proportionate. Legal safeguards should protect individuals' rights and ensure that any restrictions are based on sound scientific evidence and ethical considerations.

### 4.3.2 Ethical Decision-Making

Ethical decision-making frameworks should be integrated into public health policies. These frameworks should prioritize principles such as beneficence, nonmaleficence, and justice, ensuring that measures taken during a pandemic are ethically justified. Public health decisions should be made transparently, with input from a diverse range of stakeholders, including ethicists, healthcare providers, and community representatives.

## *4.4 Supporting Healthcare Workers*

Healthcare workers are at the forefront of the pandemic response and face significant risks and challenges. Ensuring their safety and well-being is crucial. This includes providing adequate personal protective equipment (PPE), mental health support, and clear guidelines for managing ethical dilemmas that arise during the crisis.

## *4.5 Addressing Misinformation*

Misinformation about COVID-19 poses a significant barrier to effective public health measures. Efforts should be made to combat misinformation through fact-based public health campaigns and partnerships with media organizations. Clear, consistent, and transparent communication from trusted sources is essential to counteract the spread of false information.

## *4.6 Equity and Justice*

Public health measures should be designed and implemented with a focus on equity. This includes ensuring that vulnerable and marginalized populations have access to healthcare and resources needed to comply with public health directives. Equity considerations should guide the distribution of vaccines, medical treatments, and other critical resources.

## 5 Conclusion

The Bill of Rights provides protection for individual rights which include autonomy; nevertheless, all the rights and liberties can be limited if the reason is legally justified. Therefore, the notion of a form of limited or restricted autonomy is not inconceivable, and as much it is strongly protected, it can be limited in the context presented in this chapter under strict rules to enable justice for society and without terminating the rights of the individual patients. Although the aspect of autonomy remains an important and central value in healthcare, in the face of public health emergencies, we should query if limited autonomy can be justified if it saves a human life or a community. The principles of beneficence, nonmaleficence, and justice would have to be factored in this scenario. It is important for a person, a patient, or would be patient in such circumstances to be given specific and clear information regarding the pandemic. Since most of them express a sense of fear, it would be the responsibility of the state through medical personnel and state machinery in reducing and or eliminating general public their fears and misinformation in order to provide a sense of confidence in the population. Therefore, although there is a compelling argument for promoting autonomy of patients as a means of upholding respect and dignity for patient autonomy, there is strong evidence that it is not always possible, particularly when the quality of decision and its impact on others are, to an extent, disputed.

## References


Afolabi, M. O. (2018). Inequities in health and healthcare: A persistent global issue. *The Lancet Global Health, 6*(6), e616–e617.

Aristotle. (350 BCE). *Nicomachean ethics*. Oxford University Press.

Beauchamp, T. L., & Childress, J. F. (2001). *Principles of biomedical ethics*. Oxford University Press.

Beauchamp, T. L., & Childress, J. F. (2013). *Principles of biomedical ethics*. Oxford University Press.

Caplan, A. L. (2014). Why autonomy needs help. *Journal of Medical Ethics, 40*(5), 301–302.

Center for Disease Control and Prevention. (2020). *Quarantine and isolation*. CDC.

Center for Policy Impact in Global Health. (2020). *Kenya policy response to COVID-19*. Retrieved from https://centerforpolicyimpact.org/wp-content/uploads/sites/18/2020/08/Kenya-Policy-Response-to-COVID.pdf

Chau, D. C. (2021). The geography and politics of Kenya's response to COVID-19. *PRism, 9*(4), 212–222.

Cheshmehzangi, A., Su, Z., Jin, R., Dawodu, A., Sedrez, M., Ardakani, S. P., & Zou, T. (2023). Space and social distancing in managing and preventing COVID-19 community spread: An overview. *Heliyon, 9*(3), e13879.

Childress, J. F., Faden, R. R., Gaare, R. D., Gostin, L. O., Kahn, J., Bonnie, R. J., ... & Nieburg, P. (2002). Public health ethics: mapping the terrain. *The Journal of Law, Medicine & Ethics, 30*(2), 170–178.

Dawson, A., & Verweij, M. F. (Eds.). (2007). *Ethics, prevention, and public health*. Oxford University Press.

Draper, H., & Sorell, T. (2002). Patients' responsibilities in medical ethics. *Bioethics, 16*(4), 335–352.

Engelbrecht, S. F. (2014a). Can autonomy be limited – An ethical and legal perspective in a South African context? *The Journal of Forensic Odonto-Stomatology, 32*(Suppl 1), 34–39.
Engelbrecht, S. (2014b). The right to refuse treatment in South Africa: A constitutional analysis. *South African Journal of Bioethics and Law, 7*(2), 53–56.
Epstein, R. M. (2006). Making communication research matter: what do patients notice, what do patients want, and what do patients need? *Patient Education and Counseling, 60*(3), 272–278.
Epstein, R. M., & Peters, E. (2009). Beyond information: Exploring patients' preferences. *JAMA, 302*(2), 195–197.
Etzioni, A. (1993). *The spirit of community: rights, responsibilities and the communitarian agenda*. Crown Publishers.
Gostin, L. O., & Wiley, L. F. (2020). Governmental public health powers during the COVID-19 pandemic: Stay-at-home orders, business closures, and travel restrictions. *JAMA, 323*(21), 2137–2138.
Gray, J. (1981). John Stuart Mill on liberty, utility, and rights. *NOMOS: American Society for Political and Legal Philosophy, 23*, 80.
Grogan, J. (2022). *Impact of COVID-19 measures on democracy and fundamental rights – Best practices and lessons learned in the Member States and third countries*. European Parliament.
Güner, H. R., Hasanoğlu, İ., & Aktaş, F. (2020). COVID-19: Prevention and control measures in community. *Turkish Journal of Medical Sciences, 50*(9), 571–577.
Hospital News. (2020). *Are there limits to a patient's autonomy in making health care decisions?* https://hospitalnews.com/are-there-limits-to-a-patients-autonomy-in-making-health-care-decisions/
http://kenyalaw.org/kenyalawblog/kenyas-response-to-covid-19/
https://covid19.who.int/. Retrieved May 31, 2021.
https://covid19.who.int/region/afro/country/ke. Retrieved May 31, 2021
Human Rights Watch. (2020). *Human rights dimensions of COVID-19 response*. Human Rights Watch.
Interim infection prevention and control recommendations for coronavirus disease 2019 (COVID-19) in health care settings. Ministry of Health-Kenya. https://www.health.go.ke/wp-content/uploads/2020/04/Kenya-IPC_Considerations_For-Health-Care-Settings-1.pdf
Johnstone, M. J., & Turale, S. (2014). Nurses' experiences of ethical preparedness for public health emergencies and healthcare disasters: A systematic review of qualitative evidence. *Nursing & Health Sciences, 16*(1), 67–77.
Kenya Law. (2020). *Public legal information on Kenya's response to COVID-19*. Kenya Law Blog. Retrieved from https://new.kenyalaw.org/articles/2020-04-07/kenyalawblog/public-legal-information-on-kenyas-response-to-covid-19
Meglino, B. M., & Korsgaard, A. (2004). Considering rational self-interest as a disposition: Organizational implications of other orientation. *Journal of Applied Psychology, 89*(6), 946.
Mill, J. S. (1859). *On liberty*. Broadview Press.
Mill, J. S. (1863). *Utilitarianism*. Parker, Son, and Bourn.
Mill, J. S. (1968a). *On liberty*. Penguin.
Mill, J. S. (1968b). *Utilitarianism, on liberty, essay on Bentham* (Vol. 140). World Publishing Company.
Mill, J. S. (1974). *On liberty* (G. Himmelfarb, Ed.). Penguin Books, 70, 303.1.
Mill, J. S. (1998). *On liberty and other essays*. Oxford University Press.
Ministry of Health, Kenya. (2020). *COVID-19 updates*. Ministry of Health, Kenya.
Mutahi, B. (2020, April 20). Coronavirus: The fear of being sentenced to a Kenyan quarantine centre. *BBC News*, Nairobi.
Naess, A. C., Foerde, R., & Steen, P. A. (2001). Patient autonomy in emergency medicine. *Medicine, Health are and Philosophy, 4*(1), 71–77.
Nussbaum, M. C. (2000). *Women and human development: The capabilities approach*. Cambridge University Press.
O'Neill, O. (2002). *Autonomy and trust in bioethics*. Cambridge University Press.
Rawls, J. (1971). *A theory of justice*. Harvard University Press.

Rogers, W. A. (2002a). Evidence-based medicine in practice: Limiting or facilitating patient choice? *Health Expectations, 5*(2), 95–103.
Rogers, W. A. (2002b). Evidence based medicine and justice: A framework for looking at the impact of EBM upon vulnerable or disadvantaged groups. *Journal of Medical Ethics, 28*(2), 71–75.
Rousseau, J. J. (1762). *The social contract*. Du Contrat Social.
Savulescu, J., Persson, I., & Wilkinson, D. (2020, July). Utilitarianism and the pandemic. *Bioethics, 34*(6), 620–632. https://doi.org/10.1111/bioe.12771. PMID: 32433782; PMCID: PMC7276855.
Seedhouse, D. (2008). *Ethics: The heart of health care*. Wiley.
Seedhouse, D. (2009). *Ethics: The heart of health care*. Wiley.
Stewart, K. A. (2008). Anthropological perspectives in bio-ethics. *International Encyclopedia of Public Health, 2008*, 184–193.
Sumner, L. W. (1996). *Welfare, happiness, and ethics*. Clarendon Press.
Ten Have, H. (2016). *Vulnerability: Challenging bioethics*. Routledge.
Ungerleider, R. M., Nelson, K., Cooper, D. S., Meliones, J., & Jacobs, J. (2018). *Critical heart disease in infants and children E-book*. Elsevier Health Sciences.
United Nations. (1948). *Universal declaration of human rights*. United Nations.
United Nations Human Rights Council. (1948). *Universal declaration of human rights*. United Nations.
Universal Declaration on Bioethics and Human Rights. (2005). UNESCO.
Walzer, M. (1983). *Spheres of justice*. Basic Books.
Wasike, A. (2020). *Kenya records low turnout in mass coronavirus testing*. Anadolu Agency. Retrieved from https://www.aa.com.tr/en/africa/kenya-records-low-turnout-in-mass-coronavirus-testing/1828539
WHO Coronavirus Disease (COVID-19) Dashboard. Data last updated: 29 May 2020, 7:02 pm CEST. https://covid19.who.int/?gclid=CjwKCAjw5cL2BRASEiwAENqAPhBAKxYVE5PxYQN1-Qvt_3sFs-KUbJLLWyyAu5nwoCdPAY-B9odWnRoCR4IQAvD_BwE
World Health Organization. (2020a). *Coronavirus disease (COVID-19) pandemic*. World Health Organization.
World Health Organization. (2020b). *COVID-19: Public health emergency of international concern (PHEIC). Global research and innovation forum: Towards a research roadmap*. World Health Organization.
World Health Organization. (2020c). *Ethical considerations to guide the use of digital proximity tracking technologies for COVID-19 contact tracing: interim guidance, 28 May 2020* (No. WHO/2019-nCoV/Ethics_Contact_tracing_apps/2020.1). World Health Organization.
Yusuf, M. (2020). Kenyans confined at quarantine facilities see it as sentence, April 22, 2020 03:44 pm. https://www.voanews.com/covid-19-pandemic/kenyans-confined-quarantine-facilities-see-it-sentence
Zolkefli, Y. (2017). Evaluating the concept of choice in healthcare. *The Malaysian Journal of Medical Sciences, 24*(6), 92–96. https://doi.org/10.21315/mjms2017.24.6.11

# Part II
# Climate Change and Environmental Ethics

# Chapter 4
# Environmental Ethics: A Kenyan Perspective

**Violet Naanyu, Everlyn Kisembe, Daniel Lagat, and Jane Wathuta**

**Abstract** Destruction of biodiversity results in extensive harms. The use of localized illustrative cases can illuminate the value of safeguarding the interconnected system of living and nonliving things. Being human and alive has ontological value; however, all other living things have intrinsic value too. Using literature review and illustrative cases from Kenya, we show how biodiversity is easily destroyed resulting in environmental pollution, depletion of resources, extinction of species, and climate change. More efforts are needed to create awareness on the shared natural and manmade environment, and the factors putting nature at risk. To increase sustainability and self-reliance in Kenyan communities, application of principlism ethics, participatory dialogues, and multidisciplinary work will result in safeguards for both human and nonhuman forms. Such approaches will ensure exploitation of natural resources to meet common needs while paying attention to intellectual property rights. Furthermore, increased caring between producers and consumers, participatory dialogues about the shared environment, and environmental disobedience are all vital avenues to desired change against threats noted with climate change. Otherwise, current and future generations will suffer due to our current activities.

**Keywords** Anthropocentrism · Biodiversity, Environmental Justice · Future Generations · Pollution · Sustainability

V. Naanyu (✉) · E. Kisembe
School of Arts and Social Sciences, Moi University, Kesses, Kenya
e-mail: vnaanyu@ampath.or.ke; lynnkisembe@mu.ac.ke

D. Lagat
Department of Education, School of Business and Education, Kirinyaga University, Kerugoya, Kenya
e-mail: dlagat@kyu.ac.ke

J. Wathuta
Strathmore Law School, Strathmore University, Nairobi, Kenya
e-mail: jwathuta@strathmore.edu



L. Omutoko, W. Jaoko (eds.), *Bioethics from the Global South*, Advancing Global Bioethics 21, https://doi.org/10.1007/978-3-031-77669-4_4

# 1 Introduction

Environmental ethics is the study of the moral relations between human beings and the environment. It is also concerned with the current and future status of humans and all other living creatures including plants and animals. Environmental ethics is well delineated using principlism ethics (beneficence, nonmaleficence, autonomy, and justice) because it calls for sustainable development, biodiversity preservation, the curbing of anthropogenic sources, and critical analysis of moral dilemmas at both local and global levels (Afolabi, 2018; Ten Have, 2016). In this chapter, the authors provide an environmental ethics perspective from Kenya, a nation with extensive environmental degradation, although there are several state organizations set up to support and enforce policies and guidelines governing the management of the environment. The authors ran a literature review to capture Kenya's situation about the following subtopics: environmental pollution, depletion of resources, the extinction of species, biopiracy, and environment conservation. The authors began by developing a shared understanding of environmental ethics, principlism ethics, and the narrowed subtopics of interest. Using academic databases like PubMed and Google Scholar, relevant scholarly work was identified using key terms related to the selected subtopics. The authors collected the literature and looked for common themes and trends in environmental ethics in Kenya. They arranged their findings into a logically flowing structure including the use of illustrative case studies and identified gaps where further government and scholarly work is needed. The final analysis shows the current environmental scenario, ongoing discourses, and governmental interventions and identifies opportunities for new deliberations and research on environmental ethics that would apply principlism ethics and stakeholder engagement.

Kenya is an East African country bordered by South Sudan, Ethiopia, Somalia, Tanzania, and Uganda. Kenya is undergoing extensive environmental problems, making it difficult to achieve important goals such as the Millennium Development Goals (MDGs). Removal of poverty and hunger requires that the country manages its environment effectively. These problems have been worsened by the fact that Kenya is semiarid with high temperatures and low precipitation, frequent droughts, water scarcity, and unpredictable climate variability which will have the largest impacts on people living most since it directly or indirectly supports 80% of the population and agroindustries, which support much of Kenya's economy (Kabubo-Mariara & Mwabu, 2007). According to the UNDP (2020), the mean annual temperature in Kenya has increased by 1.0 °C since 1960, representing an average rate of 0.21 °C per decade. It is projected to increase by 1.0–2.8 °C by 2060s and by the year 2100, temperatures in Kenya could increase by about 4 °C causing variability of rainfall by up to 20% and significantly compromising the ability of the Kenyan environment to rejuvenate itself.

Organizations such as the National Environmental Management Authority (NEMA), Kenya Wildlife Service, Kenya Parks Service, and the Water Regulation Management Authority exist to support and enforce policies and regulations governing environmental management in Kenya. Over the years, enforcement of various environmental policies and regulations has been wanting, a problem which exposes the existing organizations as being inept or incapable of enforcement, perhaps because of a lack of coordinating network between relevant ministries. Nonetheless, although the current legislation is quite comprehensive, some elements such as regulation on the management of air quality are missing.

Kenya's current environmental regulatory regime originates from Parliament's passage of the Environmental Management and Co-ordination Act (EMCA) of 1999. Apart from the creation of NEMA, the Act begins by granting all Kenyans the right to a "clean and healthy environment." Before the passage of EMCA, which aims to address the whole gamut of environmental regulatory issues facing a modern state, Kenya lacked comprehensive environmental regulation legislation. The EMCA of 1999 is expansive, but its most important contribution to the governance of environmental regulations is the creation of the National Environmental Management Authority (NEMA) (EMCA, 1999 part II § 7). NEMA is charged with enforcing EMCA's provisions as well as the subsidiary legislations that have been passed over the last decade. The subsidiary legislation includes water quality, waste management, controlled substances, biodiversity, wetland, river and seashore, and environmental impact assessment (EIA) regulations. Most of the provisions contained in EMCA, as well as the subsidiary legislations, are intended to provide regulations for the usage and type of allowable activity in the different ecosystems of Kenya. Thus, NEMA's main task is to review and grant licenses to proponents who plan to change land use. To complete this task, EMCA grants NEMA the power to compel any authority or ministry to comply with existing environmental regulations.

The available regulations show Kenya understands the problem of environmental dilapidation and has already spelled out what needs to be done, yet the situation remains bad. Brown (2006) opines that anthropocentrism is to blame for human insistence on unsustainable exploitation of the environment because it positions humans as consumers rather than stewards of the environment. Lane (2017) notes that humans in the modern era seek unrestrained freedom to exploit nature, and this compromises our shared environment. He argues that humans ought to cease being the "center" of the universe and be seen as part of it (Lagat, 2019).

In the following section, selected cases are used to vividly portray matters that of environmental ethics concerns in Kenya and applicable elements of principlism ethics. They cover crop and animal farming, wetlands and marine life, forestlands, mining, industrialization, and urbanization.

## 2 Case 1. Crop Farming: Reflections on Land Use Among the Bukusu

The Bukusu community is part of the Luyia people who reside in Western Kenya. The concept of ethics to the Bukusu is seen in the following phrase: "*omundu we limanya*," literally "a rational person who also has capacity to discern the right and the wrong, the good and the bad, and is inclined to act accordingly" (Makokha, 2018). To the Bukusu, a human being is a steward or a custodian, not only to others but also to nature. Land is highly valued as echoed in the following saying "*liloba lilinda omundu mubulamu nenda mwifwa,* stating that the earth sustains humanity both in life and in death (Makokha, 2018). With this kind of relationship between man and the earth, and what seems to be the nexus of identity between men, the earth, and nonhuman forms, what forms the crop farming practices among the Bukusu?

The Bukusu value land, not only for economic providence but also for social and religious inclinations, with a great reference to their ancestral land which culturally is inherited and rarely sold to a stranger. In an interview with a traditional Bukusu circumciser, he had to return back home to his grandfather's land. He says,[1]

> We returned in 1965, and my grandfather who had shared his land among his five sons, had some left over. Asked who the leftover land was for, he pointed at me saying it was for his grandson, who would care for him till death. My grandfather circumcised me in this home. I stay here because my grandfather said so.

This is a common scenario in a Bukusu home that even those who choose to buy land elsewhere will have a house built in their ancestral home and will opt to be buried there. Among the Bukusu, land was owned communally, but this is slowly changing as families buy and own land. During the precolonial period, the Bukusu grew sorghum, finger millet, simsim, black peas, green grams speckled maize, beans, cowpeas, pumpkins, sweet potatoes, bananas, and leafy vegetables (Ndalilah, 2012).

The green vegetables planted attracted weeds and other wild vegetables, e.g., *litoto* which were and are also eaten as a vegetable. The weeds which were mostly plant herbs were medicinal. Some were specially used for cultural activities such as circumcision ceremonies and/or strictly used in marriage ceremonies. During circumcision ceremonies, specific herbs are used on a circumcised boy to speed up the healing, and others used to enhance moral values. For instance, a traditional circumciser applies a painful herb to the wound to serve as a deterrent to the stubbornness of a child. The community also believes that the use of this herb ensures procreation, for it is used to open up the veins in a male's system allowing for free flow and smooth movement of the sperm during fertilization. However, many plants are disappearing and to conserve them, some people have resorted to planting them.[2] Hybrid seeds from Kenya Seed Company are reported as lacking the requisite nutritional value as the "original" seed. The farming of other foods has also drastically reduced. Sorghum for instance is a rare plant, while millet is done but on a very small scale. Ground nuts and *"njugu mawe"* were common farming crops among

---

[1] Data from the *Changing Life Project- African Identities, Moralities and Wellbeing (*AIM-WELL Project), hosted by the ACC- Moi University, Eldoret, Kenya, Grant funded by the Africa Multiple Cluster of Excellence, Bayreuth, September 2020.

[2] Personal communication with a Bukusu traditional circumciser.

communities, but are rare now. Maize was viewed as a hunger break-up crop and was not planted in bulk but is now a cash crop (Ndalila, 2012).

Farming is at the center of Bukusu activities, and it is almost a taboo to leave land fallow. The responsibility of farming among the Bukusu happens to fall on the women, who never owned land, but rather farmed on their husbands' farms or their parents'. A commonly heard phrase, "*omukhasi we ebmako,*" literally translated as "a woman with the hoe in her grip," means a hard-working woman, who understands farm production, can ensure that her family has farm produce for household consumption and trade with other communities. Through the formation of supportive women groups, some resolved to weed or harvest for each of the members taking turns to help and support one another, making farming a lighter load.

Crop farming and environmental ethics concerns among the Bukusu community reveal issues of identity, morality, and nutrition not only as symbolic to the community but also social. Green vegetables are "classified," for instance, and those used in cultural ceremonies are greatly valued, specifically for nutritional aspects but also to serve as a foundation for establishing solid and lasting social relationships. However, to increase productivity, and also in response to climate change, the seedling industry produces modified seedlings of vegetables that are perceived by the Bukusu community as carrying low nutritional levels, and which are also less familiar to the community. The green vegetables, among other local crops grown, form a strong collective identity and a symbol for social interaction for the Bukusu community. Transitioning from the use of the local seedlings to modified seedlings triggers an ethical concern as community members have to navigate changes in societal norms. Future research can consider development of modified seedlings with input from the local residents.

Within the community, one will hear some members referring to certain vegetables as *"sechili chindulu ta,"* that is "they are not bitter herbs," and *"sechikhuyeta ta sechilimo bukhala ta,"* meaning "they have no value to us." The majority of the population is aware that crucial ingredients no longer are contained in the current harvests where modified seeds are used and as such, members are resorting to other alternative means of regaining the nutritional values lost in the crops such as the use of herbs (herbal medicine). Adapting to change as a result of climate change, the community is turning to specializations of farming: the use of modified seeds due to the minimal availability of original homemade seeds, and integrated farming activities, as an effort to regain the original content of the crops, using both organic and inorganic manure, and more diverse farming activities aimed at gaining what is felt to be lost. Based on the value the community upholds on land and the farming of different crops, how can the community practice or enable integrated farming ethically?

Discussions on modified seeds are closely linked to concerns about genetically modified organisms (GMOs). Debates prevail in Kenya on GMOs and many African governments are yet to embrace GMOs—some have even banned GM crops lest they lose export markets and development aid (Wafula et al., 2012). Whenever a new GM crop is introduced, there should be transparency because it is unethical to surprise farmers with crops that bring along new costs and loss of valuable traditional underpinnings of farming. We need to have more GMO scientists engaged with farmers, consumers, and policymakers for the betterment of our shared world.

## 3 Case 2. Pastoralism Among the Maasai

Pastoralism is the keeping of domestic animals, e.g. goats, sheep, camel, and cattle. There are different forms of pastoralism depending on the ecology (Gibson, 2020) and numerous global communities practice it in arid and semiarid rangelands. Nonetheless, there is evidence from the Food and Agriculture Organization (FAO) showing the number of pastoralists has been declining over time due to increasing agriculture activities, urbanism, and also stigma directed at nomadic communities (Scanes, 2018). The Maasai of Kenya practice a mixed approach that includes two main patterns: they move along with their herds in search of pasture, and they migrate seasonally in search of fresh pastures. The animals are a main source of livelihood for the Maasai of Kenya because they provide shelter, clothing, and food. They are a ready source of milk, meat, blood, fat, and other staple food products. They also provide numerous byproducts including hide and leather, utensils made from bones and horns. They also use cowhide and dung to make beds, walls, and roofs. It is no wonder the Maasai are known to have a special affinity for their cows which have a lot of social value. A common Maasai greeting is regarding cattle: "*How are the children? How are the cattle?*" They are proud to have large herds which signify status and most rituals include the exchange or slaughter of animals.

The Maasai are known for living in harmony with wild animals and plants in their ecology. However, they are currently facing challenges due to natural and man-made risks (Gibson, 2020; Ontiri et al., 2019). The competition for pasture between wild animals and livestock is one of the big challenges noted among the Maasai because it directly affects their livelihoods. In addition, droughts have led to the loss of livestock over and over again (Ontiri et al., 2019). The grazing pastures are reducing and pastoralists are in constant competition for land as sedentary agriculturalists, animal sanctuaries, game reserves/wildlife parks, and urbanization battle for limited natural resources (Gibson, 2020). Furthermore, the Maasai rely on climatic conditions, and current unfavorable weather changes have affected the rainfall patterns and availability of pasture. Past evidence shows unanimous agreement that solutions are urgently needed to allow coexistence of pastoralists and animals because the Maasai respect wildlife in general and value tourism (Ontiri et al., 2019).

The Maasai have considered diverse ways of coping, such that, despite the water and pasture scarcity, they have adapted their systems, synergized, and evolved with the wildlife, sedentary farmers, and urbanites in their vicinity (GIAHS-FAO, 2011). There is a need to apply an ecocentric perspective when looking at ways of managing wildlife alongside Maasai cultural and livelihood sustainability (Ontiri et al., 2019; Watts et al., 2015). Successful coexistence of people, plants, and animals requires multistakeholder involvements in dialogues, decision-making, and governance of ecologies such as in Maasai land. The Maasai of Kenya uphold human rights because they value all people—from inception to old age. They also easily fold in refugees and share whatever little food and water they have with sojourners. Regarding biodiversity, the Maasai believe in safeguarding all sources of life and livelihood, e.g., they protect the pasture they graze their animals; the salt licks for all domestic and wild animals; the water points for humans, domestic, and wild animals; and the hills and caves where they perform their ceremonies. Nowadays, many pastoral lands have been turned into crop farming, industrial land, and housing estates. This has led to rapid degradation of Maasai land with consequent desertification and loss of Maasai livelihood.

## 4 Case 3. Forestlands and the Case of the Ogiek of the Mau Forest

Kenya has recorded a lot of cases where indigenous communities are battling the government over ownership, evictions, and/or use of forest lands. Communities residing in forest-related lands have commonly faced evictions justified based on conserving the environment. Such forestry governance uses guidelines and laws enforced by the Kenya Forest Service, a body that is mandated to protect all public forests. Historically (Sang, 2001), the Ogiek resided in forested areas, such as the Mau Forest, in southeastern Kenya. As a hunter-gatherer group, the Ogiek are renowned hunters, using various techniques to trap their prey. The Ogiek's diet and economy heavily relied on honey, wild fruits, and game which they got from the forests and sometimes sold to other neighboring communities in exchange for other items.

Kenya's conservation policies that barred people from living and hunting in the forest continue to marginalize the Ogiek community who have been unable to secure ownership for their forest land. They were forced by the Kenyan Government to leave their ancestral land, the Mau Forest. Their source of livelihood and social activities are dependent on the forest. Banning them from the forest endangers the community as a majority of the members are scattered and they lack access to their livelihood environment. Despite having their rightful claim to the Mau Forest territory affirmed by the African Court of Human and People's Rights ruling in 2017, the Ogiek are still unable to return to their land. Yet, Blackburn (1974) argues that it is possible for the Ogiek to remain in the area which measures only 20–40 miles in width, located on different elevation levels, ensuring a supply of different foods throughout the year. The Ogiek are "uniquely specialized people intimately related to a particular ecosystem and are incapable of retaining their characteristics if that ecosystem is destroyed" (Blackburn, 1974).

Currently, the government is pushing for the Forest Conservation and Amendment bill 2021.[3] This bill seeks to delete clause 34(2) from the 2016 act which protects forests from actions that threaten or endanger species while allowing authorities to have the constitutional right to reject a decision made by a lawmaking body or anyone who wants to alter forest boundaries. This bill has been contested by various bodies, but should it pull through, what implications will this have on the Ogiek? As a survival means, Sang (2001) observes that there is a transformation of the Ogiek into pastoralists and peasant farmers as the Ogiek continue to adopt their neighbors' ways of life. Some of the people in the Ogiek community retreat in an effort to shun away interacting and assimilating into the non-Ogiek neighboring communities so as to maintain their identity. The Ogiek are under pressure and risk losing their ancestral home and its unique features.

## 5 Case 4. The Wetlands: Nyando River Basin

Nyando River Basin is 153 km long and covers an estimated area of 3517 $km^2$. Geographically, it is located between longitudes 34°47″E and 35°44″E and latitudes 0°07″N and 0°20″S. It receives drainage from the Mau highlands, and the Nandi Hills escarpments (p. 1). What makes this ecosystem very important is that it drains its water into Lake Victoria, which is the largest freshwater lake in Africa (around 68,000 $km^2$) and which supports socioeconomically and ecologically, about 45 million people that live around it (World Bank, 2018).

One leading environmental crisis in Kenya is about water and wetlands. In many Kenyan communities, wetlands are usually viewed as "wastelands," a perspective that propagates neglect and lack of concern about the welfare of those places. It has even been claimed that during some land subdivisions, wetlands are usually not preferred, and end up being "forced" on people, in a way that each person ends up getting at least a piece. Even so, this attitude makes many people disregard the wetlands. They do not see or understand the benefits of these wetlands, which in turn prevents them from conserving them. Many wetlands have had the same treatment from the local communities. Nyando River Basin is a case example of such. Shepherd et al. (2000) claimed that this basin has been identified as the most polluted drainage basin in Kenya.

According to the United Nations Development Program (2013), Nyando River Basin has gravely been degraded over several decades. The populations around this wetland seem to have overdepended on the wetland, for meeting their basic needs. In the process, the activities on this wetland have been numerous and unsustainable. The population around this wetland is high, and as a result, there is increased damping of wastes, from homes, and other economic activity wastes, from industries upstream. Such activities place the wetlands in a more precarious position.

[3] Forest Conservation Amendment Bill. (2021). www.parliament.go.ke

According to Nyilitya et al. (2021), nitrate pollution in Nyando River Basin comes from sewage and farm chemicals (including manure) released upstream (2021, p. 1). Juma et al. (2014) assert that the urban communities and the industries around the wetland release untreated sewage, and even when attempts at treatment is made, it is usually far below standard. Commercial tea farming in Kericho and Nandi counties, together with commercial sugarcane and rice farming at Kano Plains, contributes to nitrate and phosphate leakage into the basin, and subsequently, into Lake Victoria as a result of inorganic fertilizers used. The recent increase in population around the basin has consequently led to the corresponding increase in the level of pollution and encroachment into the forests, leading to the wanton destruction of trees, which would have otherwise helped the ecosystem (Maghanga et al., 2013).

Wetlands have been converted as agricultural lands, with activities adversely compromising the quality and quantity of water available for human consumption. Droughts and unreliable rain patterns are partly to blame for this recent activity on wetlands. In other parts, wetland is a great attraction because of the fertility of the soil (most of the other available farmlands have been eroded and over ploughed, so that they are no longer fertile). The use of technology and agricultural chemicals on the wetlands has also had devastating effects on the water that flows from the wetlands and on other springs around those places. In weaker economies, governments are not able to make or enforce limiting policies, because the focus is producing food, and flowers for export, with the latter, being the priority (Lagat, 2019).

Management of Kenya Wetlands is guided by the regulation developed by the Ministry of Environment (2009). The regulation spells out the need to maintain an inventory of all wetlands in Kenya and the development management plans developed to help prevent and control degradations. It lists the actual locations of those wetlands, the types of fauna, flora, and soil; the various hydrological characteristics, and the populations depending on the wetland catchment. The regulation spells out the types of activities that may be permitted and also demands that boundaries of the wetlands need to be kept. All approved activities require a permit and an environmental impact assessment, and utilization is expected sustainably. Owners of lands or occupiers of lands adjacent to any wetland must prevent the degradation or destruction of the wetland. Failure to do so is considered an offense.

## 6 Case 5. Industrialization and Urbanization: Ban on Plastic Pollution

The Kenya Vision 2030 was launched in 2008, with the overall aim of transforming Kenya into a newly industrializing, middle-income country providing a high quality of life to all its citizens by 2030 in a clean and secure environment.[4] The Constitution of Kenya 2010 in turn provides the fundamental legal framework for the State's duty

[4] https://vision2030.go.ke/

to promote economic development while respecting the environment and sustaining it for the benefit of future generations.[5] Environmental laws and policies are among those that should be permeated with national values and principles of governance and upheld by all parties concerned. The Bill of Rights, furthermore, is projected as the framework for social, economic, and cultural policies. Article 42 states the right of every person to a clean and safe environment, and Article 72 further anticipates the enactment of legislation to safeguard the right to a safe environment. This latter provision is at the moment assured by the preexisting Environmental Management and Co-ordination Act of 1999 (Cap 387 of the Laws of Kenya).[6]

From the foregoing, it can be noted that the state has a dual and inseparable role to promote development[7] while ensuring safe environmental conditions for every person. The following case study illustrates a landmark event in Kenya's recent history and present circumstances which illustrate efforts to balance the benefits of industrialization and urbanization with the right to a clean and healthy environment. Kenya's ban against plastic pollution is presented as a salient example of the tension between industrialization and environmental protection.

In 2017, after more than a decade of engagement with stakeholders, it became illegal to use, manufacture, and import plastic bags in Kenya, whether utilized for commercial or household packaging purposes.[8] Kenya led the East African Community in boldly formulating and implementing this new law. Thereafter, in 2020, the use of single-use plastic bottles, crockery, and cutlery was prohibited in national parks,[9] further reiterating Kenya's leading role in environmental conservation. This ban was applied through the Environmental Management and Co-ordination Act, with the overall aim of protecting our biodiversity and ecosystem by drastically limiting[10] the use and improper disposal of plastic. Among the threats noted are plastic pollution in the oceans marine litter, the blocking of waterways, flooding during the rainy season, and ingestion by livestock.[11]

Manufacturers of plastic bags were encouraged to invest in the production of alternative eco-friendly bags. Contravention of this measure has already resulted in the imposition of fines or jail terms as prescribed by the Act.[12] Additional measures entailed the communication of official drop-off points so that licensed recyclers

---

[5] Preamble; Art 42, 69.

[6] http://kenyalaw.org/kl/fileadmin/pdfdownloads/Acts/EnvironmentalManagementandCo-ordinationAct_No8of1999.pdf

[7] Here used alternatively with urbanization and industrialization.

[8] Vide Gazette Notice 2334 of March 14, 2017.

[9] https://www.tourism.go.ke/wp-content/uploads/2020/05/Action-Plan-Single-Use-Plastics-Ban-in-Protected-Areas-February2020.pdf

[10] Exceptions are plastic bags used for garbage and medical waste.

[11] On the threat to livestock health, cfr. https://www.nema.go.ke/images/Docs/Mediacentre/Publication/Plastic%20waste%20in%20livestock%20rumens%20publication.pdf

[12] http://kenyalaw.org/kl/fileadmin/pdfdownloads/Acts/EnvironmentalManagementandCo-ordinationAct_No8of1999.pdf

could collect them.[13] The National Environment Management Authority (NEMA) further issued clarifications on exemptions and applications for clearance.[14] The NEMA website moreover provides responses to frequently asked questions on the plastic ban.[15] Several court petitions were filed seeking to quash the ban, citing—inter alia— the imminent job losses owing to the ban. The Environment and Lands court, however, upheld the ban in 2018.[16] The outcome provided a way forward in the journey toward industrialization, without undermining the right to a clean and healthy environment. Two years later, up to 80% compliance had been noted. Among the persisting issues, however, is the continued—albeit slowly diminishing—use of other single-use plastic materials such as bottles, as well as the insistent use of polypropylene bags whose reduced quality is harmful to the environment. Raw material for the production of eco-friendly bags[17] remains a challenge, too.

Efforts are underway, nonetheless, to establish a global legal framework to curb plastic pollution[18] and Kenya is already a signatory to the Clean Seas initiative.[19] The economic upside of this ban comprises the increased investment in plastic recycling, not to mention the cleaner and thus more aesthetic and functional environment for residents, tourists, and investors. Kenya, besides, aims to spur economic growth by fostering a sustainable blue economy, using its maritime resources to create jobs. Kenya hosted Africa's first Sustainable Blue Economy conference in 2018, with the government committing to put in place policies that harness the economic potential of the oceans and coastline. Less pollution would boost public health and marine life, thus propelling the economy forward.[20]

This case study shows how lifestyle changes brought about by industrialization and urbanization can drive the manufacture of a product that with time becomes a menace to health and societal well-being. Although the subsequent ban has been deemed a blow to the economic prosperity of the affected entrepreneurs, as prevailing environmental problems are surmounted, Kenya is still set to achieve its middle-income status while delivering on the promise of every person's quality of life in a clean and safe environment, as justly and ethically envisioned in its Vision 2030.

[13] https://www.nema.go.ke/images/Docs/Notices/Press_Statement_on_Enforcement_of_the_Ban_on_Plastic_Bags.pdf

[14] https://www.nema.go.ke/images/Docs/Notices/Statement_on_Clarifications_about_plastic_bags_ban-min.pdf

[15] https://www.nema.go.ke/images/Docs/Notices/FAQS_Editd_FINAL_2.pdf

[16] http://kenyalaw.org/caselaw/cases/view/155269

[17] Made of fabric and pulp paper.

[18] https://www.greenpeace.org/africa/en/press/48976/kenyas-fourth-plastic-ban-anniversary/

[19] https://www.cleanseas.org/

[20] https://www.unep.org/news-and-stories/story/kenya-emerges-leader-fight-against-plastic-pollution

## 7 Case 6. Mining: Gemstones and Ethics in Taita Taveta, Kenya

Kenya boasts of numerous ores and minerals including gemstones, gold, coal, iron ore, limestone, manganese, diatomite, soda ash, fluorspar, titanium, niobium, and gypsum. Many mining stakeholders are often unfamiliar with theories of equity; however, they are cognizant of diverse benefits emanating from this particular industry. The mining industry helps combat poverty among poor communities that build dependency on the mining industry to stir economic growth of the rural areas. On the contrary, mining poses several threats (Jennings, 2016). It causes reduction and degradation/contamination of ground and surface water. Exposure to dust pollution, toxic waste, and air pollution leads to ill health among mine workers, neighboring communities, and those beyond mining areas. Mining disrupts local cultures and ethnic ways of life and their operations create social dislocations through work migration, a situation that would greatly benefit from a global bioethics lens (Ten Have, 2016). Sanitary-environmental problems occur as large numbers of workers arrive and settle near mines. In nations with weak laws and policies to guide mining activities, there is poor implementation of existing regulations on mining, resulting in problems such as pollution and deforestation, and they all affect the quality of life locally and beyond mine areas. Mining harms the general ecosystem and leads to changes in land use, e.g., loss of grazing lands due to erosion and pollution.

An illustrative example of mining and environmental ethics has been reported from Taita Taveta County, Kenya, where world-renowned gemstones are mined including emeralds, ruby, sapphire, garnets, and tsavorites. While the stones fetch a lot of income abroad, the source communities in Taita Taveta have not enjoyed new development or seen efforts toward alleviation of poverty. A recent study (Mwakesi et al., 2020) explored how mining activities had affected local communities' livelihoods in Mwatate Sub-county, Taita Taveta County, Kenya. The study reported weak mining regulatory systems and the limited influx of migrants into the mining areas. The mining provided daily bread for many workers; however, these earnings were minimal. Furthermore, the mining company did not rehabilitate the mining areas, and there was an inequitable distribution of benefits—some mining companies failed to compensate locals and there was negligible support for development projects. The study recommended environmental and social impact assessments and deliberate implementation of legal frameworks to regulate mining operations in Taita Taveta. This would ensure fair distribution of costs and benefits emanating from the mining activities and hopefully improve livelihoods.

As we embrace the benefits of the mining industry to the collective, we must reflect on reasonably acceptable risks because when unethically managed, mining activities result in the extinction of plant species and the depletion of natural resources. There should also be protection of cultural rights and livelihoods. Families who lose their land to mining ought to be provided commensurate arrangements. There should be more community-engaged participatory approaches in making situational ethical decisions on the mines, mining, and associated ethical conflicts. There should be laws protecting natural resources, the environment, land ownership, and indigenous ways of life. Enforcement of laws on mineral resources will make mineral license holders comply with environmental laws and restrict exploration and mining activities. We need legislation addressing the social and health impacts of mining and associated compensations. Clear guidelines are also needed on how equity will be assessed and how negative impacts on the public will be measured and minimized. It is especially important to determine if any local populations are particularly vulnerable and merit special consideration.

## 8 Efforts to Conserve the Environment in Kenya

> You cannot protect the environment unless you empower people, you inform them, and you help them understand that these resources are their own, that they must protect them.
> Wangari Maathai

In Kenya, the first legislation to conserve forests, commonly referred to as a singular control command, came into operation in 1891 (Mbuvi et al., 2018). This was later changed and in 1997, Kenya rolled out a participatory forest management pilot project. This was later institutionalized as an alternative approach that involves "Forest Adjacent Communities (FAC) and other stakeholders in sustainable management of forests within a framework whose guidelines[21] were clearly stated." These guidelines were tagged as not compulsory while carrying a note stating that they were not Kenya Forest Service regulations. These guidelines, though partly legally binding, were flexible on the extent to which any stakeholder put efforts into conserving forests; whether they were government officers, religious groups, community-based organizations, private sector investors, general community members, or individuals. They all generated ethical considerations on what is right or wrong on conservation by what each organization and/or individual view(s) or deem(s) to be right and/or wrong. The guidelines center on community identification, verifying the resource, and preparation for approval of a Participatory Forest Approach plan among others, with a focus on contributing to the community's livelihoods. These are all steps that require an objective eye and serious ethical considerations, for example, power relations. Within the participatory approach, who takes the leading role, the government or the community? This is a question that may evoke issues of power and it will arise whether we are speaking about mining, farming, or industrialization.

The participation of various organizations in conservation activities paves the way for conflicting ideas on what the best conservation approach would be. Taking on the view that all organizations are made up of "homogenous" groups of people with similar minds/thinking, the language spoken matters—competence in the indigenous languages correlates with the preservation of the environment because

[21] www.kenyaforestservice.org/documents/pfm/PFM/%20Guilines%20Final%202016.pdf

the particular language one speaks influences the way one thinks about reality (Lucy, 2001). A UNESCO report reveals that "Indigenous peoples derive their identities, values, and knowledge systems from their interaction with their territories, whether forests or seas. Their languages are shaped by their environment – they attempt to describe their surroundings and form the basis of their unique tongues. When the territory is altered, changes also occur in the culture and eventually, in the language."[22] Thus, efforts in conservation call for an examination of the complex interactions of various players and factors, and how these factors reveal ethical considerations, all for the sustainable management of the environment. These approaches ought to be encouraged for they ensure environmental justice as all stakeholders are fairly treated and meaningfully involved regardless of their social and economic status.

## 9 Biopiracy, Patenting, and Human Rights

Patent activity linking genetic resources and traditional knowledge (TK) has long been a global cause of concern due to the misappropriation of genetic resources and TK, often known as biopiracy.[23] Biopiracy has become one of sources of conflict between intellectual property (IP) rights on TK and enjoyment of fundamental human rights. A case at hand is the 2003 case of the Mweri plant whereby the government was entangled in a dispute with J. Leakey over the export of items from a tree known as "*mweri*" in Gikuyu. Mweri, also known as the African Plum (*Prunus africana*), is used to cure a variety of ailments, including genitourinary problems, allergies, inflammations, and benign prostatic hyperplasia (Oduor, 2010). Despite its well-known indigenous roots, *mweri* bark extract is trademarked by Dr. Jacques Debat, a French entrepreneur. Every year, around 300 tons of bark extract are shipped, and it costs US$2 per kilogram in Kenya, whereas capsules containing the bark extract cost around US$8 for a 15-capsule packet in Europe. As a result, the French pay Kenyans $2 to make $800, and if Kenya develops the capacity to make medications from this bark, it will be compelled to pay royalties to the patent holder (Oduor, 2010).

Governments in developing nations have started recognizing the need to conserve TK and ensure that the rights of indigenous and local community members who discover medical knowledge are recognized and protected. Moreover, increasing consideration of global bioethics provides an opportunity for continuous interactions and exchanges between local and global communities as we deliberate on global justice issues, and this is a much-needed approach toward recognition of geographic vulnerabilities, cultural diversity, and rights embedded therein (Afolabi, 2018; Ten Have, 2016).

---

[22] Indigenous Languages: Knowledge and Hope. The UNESCO Courier. e -ISSN 2220 2298.

[23] P. Oldham et al., 'Biodiversity in the Patent System: Kenya,' (One World Analytics,2013) 1.

In Kenya, TK holders and indigenous people have started using IP laws to protect their rights. Kenya is a member of the World IP Organization and has implemented a variety of IP legislation and is a signatory to the Convention on Biological Diversity and the related Nagoya Protocol on Access to Genetic Resources, implemented through the Environmental Management and Coordination Act. Furthermore, the Constitution and the Protection of TK and Cultural Expressions Act both have clauses addressing TK.[24] Kenyan institutions whose work is relevant to TK protection include the Kenya Industrial Property Institute, which administers trademarks and patents; the Kenya Copyright Board, which focuses on TK and all matters of copyright and related rights in Kenya; and the Kenya Plant Health Inspectorate Service, which administers plant protection, and plant breeders' rights (Kwamboka, 2017)

The IP regime, however, is not sufficient for protection of TK. First, IP rights fail to recognize customary laws and systems used by TK holders. Second, IP often grants exclusive ownership rights to the author or invention, which fundamentally violates the nature of TK. As a result, utilizing IP models to safeguard TK would undermine and destroy TK holders' culture and history, as well as cause biopiracy. Third, TK is transgenerational, making it difficult to identify a creator or inventor to recognize (Kariuki, 2019). There is therefore the need for more progressive and enhanced solutions for issues concerning the protection of TK holders and their rights in Kenya.

## 10 Conclusion

Environmental ethics implies rights, duties, and responsibilities to be observed even as we develop the natural resources within our localities. The application of principlism ethics in real-life illustrative case studies allows a better discussion on environmental issues. It facilitates dialogues on human rights because people need sustainable milieus established based on respectful consideration of all living and nonliving things. Within a balanced environment, our aspirations for basic rights can be assured. To avoid being unjust, as we deliberate on agricultural, economic, industrial, and associated development, concerned stakeholders ought to consider all citizenry, but more so, the neediest. Equity then becomes a key concept because as seen in mining in Taita Taveta and Ogiek forestlands, the distribution of resources should be aimed at addressing inequalities. Furthermore, as we embrace emergent technologies and industrialization, IP rights ought to be defended so that requisite recognition of customary systems and ownership rights are safeguarded.

Civic education can empower stakeholders to make informed decisions about their environment, socioeconomic context, and livelihoods. Ongoing participatory dialogues are needed on "acceptable risks," criteria for risk management and reduction, and ways to enhance indigenous knowledge and its contribution to the

[24] Constitution of Kenya, Art 11.

environment. The Bukusu case raises concerns about ethical procedures needed in seed development to ensure the maintenance of nutrition, socioeconomic fabric, and broader physical environment. Food has unique societal and historical connotation, and technological modification of food may have negative impact especially where there is no informed consent, community engagement, or room for self-determination among local populations. People should be well informed and have control over what they grow and eat and values therein.

Failure to observe environmental ethics results in a state of apathy and disgruntled citizens, and this can stimulate environmental disobedience as local populations take up actions to protect their environment from damaging activities and to prevent further destruction to their already damaged environment. We must encourage positive discourse and also have equity-focused environmental impact assessments so that consensus on these issues can be reached amicably and morally.

**Acknowledgment** The authors acknowledge the research assistance of Mitchel Kakai Namalwa, intern, Strathmore Institute for Family Studies and Ethics and Law student, Kenyatta University School of Law.

## References

Afolabi, M. O. (2018). Public health disasters. *Public Health Disasters: A Global Ethical Framework, 28*(2), 1–24.

Blackburn, R. H. (1974). The Okiek and their history. *Azania Archeological Research in Africa, 9*(1), 139–157.

Brown, E. (2006). *Our father's world.* Doorlight.

Environmental Management and Coordination (Wetlands, River Banks, Lake Shores and Sea Shore Management) Regulations. (2009). https://www.nema.go.ke/images/Docs/Regulations/Wetlands%20regulations-1.pdf

GIAHS-FAO. (2011). Oldonyonokie/Olkeri Maasai Pastoralist Heritage, Kenya. https://www.fao.org/giahs/giahsaroundtheworld/designated-sites/africa/oldonyonokieolkeri-maasai-pastoralist-heritage/

Gibson, M. (2020). *Agriculture, forestry, and fisheries.* In M. Gibson (Ed.), *Food and society* (pp. 165–178). Academic Press.

Jennings, B. (2016). Environmental and occupational public health. In L. W. Ortmann, D. H. Barrett, C. Saenz, R. G. Bernheim, A. Dawson, J. A. Valentine, & A. Reis (Eds.), *Public health ethics: Global cases, practice, and context* (pp. 177–202). Springer.

Juma, D. W., Wang, H., & Li, F. (2014). Impacts of population growth and economic development on water quality of a lake: case study of Lake Victoria Kenya water. *Environmental Science and Pollution Research, 21*(8), 5737–5746. Retrieved from https://doi.org/10.1007/s11356-014-2524-5

Kabubo-Mariara, J., & Mwabu, D. K. (2007). Determinants of school enrollment and education attainment: Empirical evidence from Kenya. *Sage, 75*(3), 575–593.

Kariuki, F. (2019). *Protecting traditional knowledge in Kenya: traditional justice systems as appropriate Sui generis systems* (WIPO-WTO Colloquium Papers), p. 94

Kwamboka, N. R. N. (2017). *Biopiracy and the case for traditional medicine in Kenya.* Strathmore Law School, Strathmore University, Nairobi. https://core.ac.uk/download/pdf/84905438.pdf

Lagat, D. K. (2019). *Christian faith and environmental stewardship.* Wipf and Stock.

Lane, D. A. (2017). Anthropological and theological reflections on Laudato Si'. In E. M. Sean (Ed.), *Laudato Si' An Irish response: Essays on the Pope's letter on the environment* (pp. 31–54). Veritas.

Lucy, J. A. (2001). *International encyclopedia of the social & behavioral sciences* (pp. 13486–13490).

Maghanga, J. K., Kituyi, J., Kisinyo, P., & Ng'etich, W. K. (2013). Impact of nitrogen fertilizer applications on surface water nitrate levels within a Kenyan tea plantation. *Journal of Chemistry, 1–4*, 196516. https://doi.org/10.1155/2013/196516

Makokha, K. (2018). Reflections on Bukusu ecological wisdom and virtue in environmental ethics discourse. *International Journal of African Society Culture and Tradition, 6*(1), 15–27.

Mbuvi, M., Ndalilo, L., & Hussein, A. (2018). Applying sustainability and ethics in forest management and community livelihoods: A case study from Arabuko Sokoke Forest, Kenya. *Open Journal of Forestry, 8*, 532–552.

Mwakesi, I., Wahome, R., & Ichang'i, D. (2020). Mining impact on communities' livelihoods: A case study of Taita Taveta County, Kenya. *AIMS Environmental Science, 7*(3), 286–301.

Ndalilah, J. W. (2012). Labour dynamics in Kenya's pre-capitalist Bukusu economy and society. *Journal of Emerging Trends in Educational Research and Policy Studies (JETERAPS), 3*(5), 646–654.

Nyilitya, B., Mureithi, S., Bauters, M., & Boeckx, P. (2021). Nitrate source apportionment in the complex Nyando tropical river basin in Kenya. *Journal of Hydrology, 594*, 1–13. https://doi.org/10.1016/j.jhydrol.2020.125926

Oduor, P. (2010). *Bio piracy, the intellectual property regime and livelihoods in Africa* (Ritimo, 2010). https://www.ritimo.org/Biopiracy-the-intellectual-property-regime-and-livelihoods-in-Africa. Accessed 25 Apr 2022

Ontiri, E. M., Odino, M., & Kasanga, A. (2019). Maasai pastoralists kill lions in retaliation for depredation of livestock by lions. *People and Nature, 1*(1), 59–69.

Sang, J. (2001). *The Ogiek in Mau Forest*. https://www.academia.edu/36235838/Case_study_3_Kenya_The_Ogiek_in_Mau_Forest

Scanes, C. G. (2018). The neolithic revolution, animal domestication, and early forms of animal agriculture. In C. G. Scares & R. S. Toukhsati (Eds.), *Animals and human society* (pp. 103–131). Academic Press.

Shepherd, G., Bureh, R., & Gregory, P. J. (2000). Land use affects the distribution of soil inorganic nitrogen in smallholder production systems in Kenya. *Biology and Fertility of Soils, 31*, 348–355.

Ten Have, H. (2016). *Global bioethics: An introduction*. Routledge.

United Nations Development Program. (2020, November 25). *Transitioning from the MDGs to the SDGs*. Retrieved from UNDP: https://www.undp.org/content/undp/en/home/librarypage/sustainable-development-goals/transitioning-from-the-mdgs-to-the-sdgs.html

United Nations Development Program (UNDP). (2013). *River Nyando Wetlands Conservation Project What the project is about*. United Development Program. https://www.ke.undp.org/content/kenya/en/home/operations/projects/environment_and_energy/river_nyando_wetlads.html#:~:text=Whereas%20it%20is%20common%20knowledge%20that%20the%20Nyando,minimize%20the%20escalating%20degradation%20of%20these%20biodiversity-r

Wafula, D., Waithaka, M., Komen, J., & Karembu, M. (2012). Biosafety legislation and biotechnology development gains momentum in Africa. *GM Crops Food, 3*(1), 72–77.

Watts, P., Custer, P., Yi, Z., Ontiri, E., & Pajaro, P. (2015). A Yin-Yang approach to education policy regarding health and the environment: Early-careerists' image of the future and priority programmes. *Natural Resources Forum, 39*, 202–213.

World Bank. (2018). *Lake Victoria Environmental Management Project: Project Information Document/Integrated Safeguards Data Sheet (PID/ISDS)*. World Bank.

# Chapter 5
# Bioethics and One Health: Beyond Anthropocentrism

**Karori Mbũgua**

**Abstract** Interest in the One Health concept has been reawakened by the recent COVID-19 pandemic and the emergence and reemergence of zoonotic diseases in general. One Health is the notion that human health is intricately tied to animal and environmental health. Unfortunately, despite its promise to adopt an integrative and collaborative approach to health and disease, the One Health approach retains an anthropocentric bias and does not adequately address the nonhuman aspects of the triad. Mainstream bioethicists are not helping the situation either. Their approach to bioethics is equally human centered. Indeed, they have little to say about how we should treat animals or the environment. The overarching objective of this chapter is to highlight these anthropocentric biases in the One Health approach and mainstream bioethics, a topic that has not been sufficiently addressed in bioethics literature. Using a conceptual approach, the chapter uncovers these assumptions and argues that both bioethics and the One Health approach could benefit from Aldo Leopold's ecocentric thought and the ecological wisdom embedded in traditional African cultures and worldviews. Indeed, incorporating diverse cultural viewpoints without compromising core moral principles will make bioethics more truly global.

**Keywords** African ecological wisdom · Anthropocentrism · Aldo Leopold · Bioethics · COVID-19 · Ecocentrism · One Health

K. Mbũgua (✉)
Department of Philosophy and Religious Studies, University of Nairobi, Nairobi, Kenya
e-mail: karori.mbugua@uonbi.ac.ke



L. Omutoko, W. Jaoko (eds.), *Bioethics from the Global South*, Advancing Global Bioethics 21, https://doi.org/10.1007/978-3-031-77669-4_5

## 1 Introduction

The recent COVID-19 pandemic underscored the value of embracing the One Health approach. It showed how interference with nature can negatively impact human health and destabilize the world, both socially and economically. Indeed, the pandemic not only sickened and killed people, but it also resulted in job losses and interfered with interpersonal and community interactions. The One Health concept is based on the premise that human, animal, and environmental health are intimately connected. This concept entails making a paradigm shift from a human-centered approach to health to an integrative and collaborative approach. The One Health strategy is considered the most efficient and promising way of preventing and controlling diseases, especially those that are passed from animals to humans and vice versa. It has been embraced and endorsed by the World Health Organization (WHO), the Food and Agricultural Organization (FAO), the World Organization of Animal Health (OIE), and the World Bank (Lee & Brummel, 2013). However, despite its worldwide acceptance, the One Health approach suffers from anthropocentric biases, and some critics have said that it is not progressive enough (Kamenshikova et al., 2021; Coghlan et al., 2021).

Mainstream bioethics itself is not immune from charges of anthropocentrism. It has primarily been concerned with human life and well-being. Although bioethics is supposed to include medical ethics, animal ethics, and environmental ethics, bioethicists have tended to focus on the medical ethics aspect of the triad, and animals and the environment are only mentioned in passing (Chaffee, 2017).[1] As Richie (2016) has pointed out, although environmental bioethics forms an important part of bioethics, the former is currently alienated from modern bioethics (Ritchie, 2016).

The main objective of this chapter is to examine the anthropocentric biases inherent in mainstream bioethics and One Health, a topic often overlooked in bioethics literature. Using a conceptual approach, I uncover these biases and argue that incorporating Aldo Leopold's ecocentric perspectives and the ecological wisdom embedded in traditional African cultures and worldviews could enhance both One Health and bioethics.

## 2 The One Health Concept

Although the term One Health was only used first in 2004 following the outbreak of severe acute respiratory disease (SARS) in 2003, the concept itself is as old as Western medicine. Since the time of Hippocrates, scientists and medical practitioners have been aware of the close connection between health and the environment. In his famous book titled *Airs, Waters and Places*, Hippocrates himself maintained

[1] It is important to emphasize that medical ethics is an umbrella term that encompasses clinical ethics, public health ethics, and research ethics, among other areas.

that good human health depended on a clean environment (Hippocrates, 1923). However, more recent roots of One Health can be traced to the One Medicine approach, which was developed in the 1960s and 1970s to integrate human and veterinary medicine (Schwabe, 1984).

One Health recognizes that most of the emerging infectious diseases facing mankind today go beyond species boundaries. Indeed, the COVID-19 pandemic brought to the fore the deep connections between human, animal, and environmental health. With improved means of travel, coupled with deforestation and diversity loss, humans and animals are interacting more freely and frequently, which offers an opportunity for pathogens to spill over from animals to humans and vice versa. The One Health initiative seeks to improve the health of humans, animals, and the environment through multidisciplinary collaborations locally, internationally, and globally (American Veterinary Medical Association, 2008).

## 3 The Roots of Anthropocentrism

Despite its widespread acceptance and its commitment to an integrative approach to health, the One Health approach retains some anthropocentric tendencies (Kamenshikova et al., 2021) and has been criticized for not being radical enough (Coghlan et al., 2021). But what exactly is anthropocentrism?

Anthropocentrism is the view that only human beings have intrinsic moral worth and only they deserve strong ethical protection. In this view, the values that other beings possess are instrumental and are assigned to them by humans. Anthropocentrism is contrasted with biocentrism, which assigns intrinsic value to all living things, and ecocentrism, which focuses on the entire cosmos and extends moral concern to nonhuman nature, both living and nonliving. Peter Singer, the world's foremost proponent of animal rights, maintains that the anthropocentric privileging of Homo sapiens is arbitrary and is as unjustifiable as sexism and racism (Singer, 1990). The anthropocentric worldview is captured in the dictum "man is the measure of all things," a statement uttered by the ancient Greek Sophist Protagoras. This statement is perhaps the most profound expression of anthropocentrism.

The roots of modern-day anthropocentrism have been traced back to the Judeo-Christian tradition. However, some form of anthropocentrism also appears in the writings of Aristotle, who maintained that there is a hierarchical arrangement of beings in the world, with man occupying the apex position (Aristotle, 1989). In his oft-cited paper, Lynn White Jr. argued that Christianity was the most anthropocentric religion and blamed it for the environmental problems facing man (White, 1967). He maintained that the Judeo-Christian tradition had placed man on the pedestal of cosmological privilege, separating him from the rest of creation and making him superior to them. The following verse best exemplifies this view:

> And God blessed them. And God said to them, "Be fruitful and multiply and fill the earth and subdue it and have dominion over the fish of the sea and over the birds of the heavens and over every living thing that moves on the earth." (English Standard Version Bible, 2001, Gen. 1:28)

According to White, this scriptural passage marked the genesis of environmental degradation. It placed man at the center of the universe with a mandate to conquer and dominate nature. It is worth mentioning that the human impact on the environment has been so far-reaching that a new geologic epoch, the Anthropocene, has been proposed to supplant the Holocene epoch, which began about 11,700 years ago (Carrington, 2016).

## 4 Anthropocentrism and One Health

As already noted, despite its worldwide acceptance, One Health still suffers from anthropocentric biases. Indeed, a recurring criticism of One Health is that its policies are primarily aimed at safeguarding human health. The promotion of animal and environmental health is regarded merely as a means of achieving that end.

An analysis of 23 international antimicrobial resistance (AMR) policy documents revealed that One Health discourses were framed in an anthropocentric language. The study found that most of the documents were primarily dedicated to improving human health, and when both animal and environmental health were mentioned, it was often in terms of their instrumental value to human health. Moreover, concerning the use of antibiotics in clinical and agricultural settings, humans were viewed as the sufferers of the disease, whereas animals and the environment were viewed as the causes (Kamenshikova et al., 2021).

Another study carried out in the Netherlands exploring the ethical presuppositions of Dutch health workers involved in zoonotic diseases prevention and control policies revealed that although the majority of the respondents endorsed the holistic idea of One Health, in practice they viewed the control of zoonotic diseases as primarily aimed at protecting and promoting human health. Besides, none of the respondents in the study said that humans and animals have the same moral status. The researcher noted that this finding significantly differed from the views of the Dutch general public, some of whom believe that humans and animals deserve the same moral consideration because they are sentient beings (Herten et al., 2020).

The continued practice of intensive farming even in countries that have embraced the One Health approach is another indication of how deeply rooted anthropocentrism is in modern culture. Although intensive farming is efficient in terms of food production, it comes with a high price for animals and the environment. The practice causes immense suffering to animals and contributes to biodiversity loss and climate change (Hampton et al., 2021). It is for this reason that Coghlan and his coworkers have called for the expansion of plant-based foods and the introduction of cultured meat (Coghlan et al., 2021).

Furthermore, despite the adoption of the One Health approach, the culling of healthy animals continues to be used as a zoonotic disease control method in many countries. Unfortunately, public health policies based on culling are inherently anthropocentric because they fail to consider the welfare of animals. A case in point is the ongoing culling of badgers in the United Kingdom to control the spread of bovine tuberculosis (bTB). The interests of farmers seem to have taken precedence over the interests of badgers. The cull was carried out without sufficient evidence of its effectiveness or moral justification (see Lederman et al., 2021). Besides, most of the approaches used when culling cause immense pain and suffering to animals. Vaccinating the badgers or the cows themselves would be a more morally acceptable policy for controlling bTB than killing the healthy badgers. Indeed, a recent study has reported that vaccination is a more effective method of controlling the spread of bTB compared to culling (Woodroffe et al., 2024).

The COVID-19 pandemic offered another illustration of how deep-rooted the anthropocentric value system is in society. Despite early confirmation that human activities such as habitat destruction and hunting of wild animals were the root causes of the COVID-19 outbreak, blame was shifted away from humans to nonhuman nature, which was seen as an enemy to be tamed and punished. In Denmark, for example, 17 million minks were killed following reports of mink-specific coronavirus mutations in humans (Frutos & Devaux, 2020). Similarly, hundreds of bats were culled in Indonesia on suspicion that they carried the COVID virus (Tsang, 2020).

## 5 Mainstream Bioethics and Anthropocentrism

Generally speaking, bioethics is comprised of three subdisciplines, namely, medical ethics, environmental ethics, and animal ethics. However, when people talk of bioethics, they usually mean medical ethics, and the other aspects of the triad are often ignored. Indeed, bioethics is often perceived to be synonymous with medical ethics.

Mainstream bioethics is guilty of misplaced anthropocentrism. It has tended to fix its attention on medical ethics at the expense of the other aspects of the bioethics triad. Like the One Health approach, mainstream bioethics continues to value human interests above nonhuman interests. Indeed, the themes that concern mainstream bioethics are those that touch on human participants and patients, while animals and the environment are only mentioned in passing. As Resnik aptly puts it, "Bioethics has traditionally focused on dilemmas and decisions related to interactions among patients, medical professionals, and healthcare organizations. Topics such as abortion, euthanasia, informed consent, privacy, reproductive health, and access to health care are the bread and butter of bioethics research, education, and consultation" (Resnik, 2009, p. 1). A survey of bioethics conference programs and papers published in bioethics journals would seem to confirm Resnik's claims (see Rawlinson, 2015).

Interestingly, Rennesealar Potter's original concept of bioethics had anthropocentric connotations. In the preface to his often-cited book entitled *Bioethics: A Bridge to the Future*, he explicitly stated that the aim of writing that book was to "*contribute to the future of the human species*" (Potter, 1971, p. vii). His understanding of bioethics differed significantly from that of Fritz Jahr, who is credited with first coining the term bioethics (Jahr, 1927). In his definition of bioethics, Jahr sought to expand the circle of moral regard to include nonhumans, emphasizing the interdependence of all living things. Later, after realizing that his definition of bioethics was rather narrow, Potter himself modified his original term and called it global bioethics.

The anthropocentric underpinnings of bioethics are also clearly evident in the Universal Declaration on Bioethics and Human Rights (UNESCO, 2005). The declaration is framed in a manner that suggests that human interests are the only ones with moral significance. Article 2(b) of the declaration explicitly states that the aim of the declaration is "to promote respect for *human dignity* and protect *human* rights, by ensuring respect for the life of *human beings,* and fundamental freedoms, consistent with international law" (emphasis mine). The word animal is only mentioned once in the declaration, and where the environment is mentioned, it is only mentioned because it benefits the human race. This is ironic considering that an International Bioethics Committee (IBC) meeting held in 2005 to discuss the content of the declaration had agreed that the declaration would focus on not just human beings but all living things (MacPherson, 2007).

Bioethicists' fixation with medical ethics is not entirely surprising; however, modern bioethics emerged as a response to several medical controversies involving humans in both Europe and America, including the Nazi atrocities in Germany during the Second World War and the infamous Tuskegee syphilis experiments in the United States, which started in the 1940s and terminated in 1970, and advances in biomedical technology in the 1960s. Specialists in the three aspects of the bioethics triad are locked in disciplinary silos, with very little communication among themselves. The well-being of humans, animals, and the environment will be best served by breaking these disciplinary silos within bioethics. To be in line with the One Health initiative, the mandate of bioethics must be expanded to include greater collaboration between all the players in the bioethics triad, namely environmental ethicists, medical ethicists, and animal ethicists.

## 6 Toward an Ecocentric Approach

Bioethics needs to move away from its current anthropocentric orientation toward a more ecocentric approach. This new orientation enlarges the circle of moral concern to include nonhuman animals and the environment. The good news is that we will not be starting from scratch. The broad outlines of ecocentrism were sketched by the American ecologist, Aldo Leopold, in the mid-twentieth century in his famous essay titled "The Land Ethic." Ecocentrism is also a defining characteristic of the traditional African worldview, which emphasizes the interrelatedness of everything in nature.

Leopold's ecocentric ethic "changes the role of homo sapiens from conqueror of the land community to plain member and citizen of it" (Leopold, 1949, p. 204). This ethic expands the moral community to include not only nonhuman animals and plants but also the nonliving environment. This paradigm shift requires that we assign intrinsic value to nonhuman animals and the environment. Following Darwin, Leopold maintained that a lack of concern for nature was a sign of deficiency in moral development. His rejection of humans as the moral center of the universe has rightly been likened to the Copernican revolution in astronomy, which displaced man from the center of the cosmos (Proctor, 1992), and just like the Copernican model, which took time to be accepted, Leopold's ecocentric ethics may take time to be fully accepted.

Mainstream bioethics and the One Health approach could also benefit from the African traditional worldview, a worldview that recognizes the interconnectedness and interdependence of all aspects of life and the need to live in harmony with nature in order to remain healthy in a holistic sense. Through a complex system of beliefs involving totems and taboos, traditional Africans had a way of extending the circle of moral concern to accommodate nonhuman nature (Mangena, 2013; Chemhuru & Masaka, 2010; Masaka, 2019). Indeed, contrary to the received view, far from being anthropocentric, African ethics is community centered. The community here is to be understood in a broad sense to encompass not just humans but all of nature (Etieyibo, 2017). This view is similar to Aldo Leopold's notion of a biotic community. As Tosam put it, "African environmental ethics extends the moral community beyond anthropocentric concerns by including nonhuman animals, plants and the unborn, and the supernatural into the moral universe" (Tosam, 2019, p. 172). Also included in this moral universe are inanimate things such as mountains and rivers because they are part of the web of life (Behrens, 2014). To be truly global, bioethics must acknowledge cultural realities while at the same time upholding universal moral values, including autonomy, beneficence, nonmaleficence, and justice.

With our planet having already entered the Anthropocene, an age characterized by intense human destruction of the environment, Leopold's insights on how mankind should relate to nature are more relevant than ever before. A retrieval of traditional African ecological wisdom and practices might prove useful in addressing the emerging health problems at the human–animal–environment nexus, such as the recent COVID-19 pandemic.

## 7 Conclusion

We have seen that despite attempts to embrace an integrated approach to health, One Health still suffers from anthropocentric bias. Human health is still given priority, and the health of nonhuman animals and the environment is only valued in terms of immediate human utility. Some of the One Health strategies, such as culling, pay very little attention to animal suffering, and most One Health policy documents are couched in the language of anthropocentrism. What the One Health approach requires is an ecocentric moral framework, one that extends moral concern to nonhuman nature. And to be in line with the One Health approach, bioethics itself will need to appreciate the connection between medical ethics, animal ethics, and environmental ethics. Workers in the three subfields of bioethics need to break out from their disciplinary silos and embrace a collaborative approach to ethical issues. Further research should focus on how traditional African ecological knowledge could be integrated into the One Health approach.

## References

American Veterinary Medical Association. (2008). *One health, a new professional imperative. One health initiative task force: Final report*. July 15, 2008. http://www.avma.org/onehealth/onehealth_final.pdf. Accessed 2 May 2022.

Aristotle. (1989). Animals and slavery. In T. Regan & P. Singer (Eds.), *Animal rights and human obligations* (2nd ed.). Prentice-Hall.

Behrens, K. (2014). An African relational environmentalism and moral considerability. *Environmental Ethics, 36*(1), 63–82.

Carrington, D. (2016). The Anthropocene epoch: Scientists declare dawn of human-influenced age. *The Guardian*, August 29.

Chaffee, M. (2017). The case for integrating the environment into the definition of bioethics. *Journal of Environmental Law, 35*(2), 255–283.

Chemhuru, M., & Masaka, D. (2010). Taboos as sources of Shona people's environmental ethics. *Journal of Sustainable Development in Africa, 12*(7), 121–133.

Coghlan, S., Coghlan, B. J., Capon, A., & Singer, P. (2021). A Bolder one health: Expanding the moral circle to optimize health for all. *One Health Outlook, 3*(1), 21. https://doi.org/10.1186/s42522-021-00053-8

English Standard Version Bible. (2001). *ESV Online*. https://esv.literalword.com/

Etieyibo, E. (2017). Anthropocentrism, African metaphysical worldview, and animal practices: A reply to Kai Horsthemeke. *Journal of Animal Ethics, 7*(2), 145–162.

Frutos, R., & Devaux, C. A. (2020). Mass culling of minks to protect the COVID-19 vaccines: Is it rational? *New Microbes and New Infections, 38*, 100816. https://doi.org/10.1016/j.nmni.2020.100816

Hampton, J., Hydman, T., Allen, B., & Fischer, B. (2021). Animal harms and food production: Informing ethical choices. *Animals, 11*(5), 1225. https://doi.org/10.3390/ani11051225

Herten, J., Buikstra, J., Bovenkerk, B., & Stassen, E. (2020). Ethical decision-making in zoonotic disease control: How do One Health strategies function in the Netherlands? *Journal of Agricultural and Environmental Ethics, 3*, 239–259.

Hippocrates. (1923). *Airs, waters, places*. Harvard University Press.

Jahr, F. (1927). Bio-Ethik: Eine Umschau über die ethischen Beziehungen des Menschen zu Tier und Pflanze. *Kosmos, 24*(1), 2–4.

Kamenshikova, A., Wolffs, P. F. G., Hoebe, C. J. P. A., & Horstman, K. (2021). Anthropocentric framings of one health: An analysis of international antimicrobial resistance policy documents. *Critical Public Health, 31*(3), 306–315. https://doi.org/10.1080/09581596.2019.1684442

Lederman, Z., Magalhães-Sant'Ana, M., & Voo, T. C. (2021). Stamping out animal culling: From anthropocentrism to one health ethics. *J Agric Environ Ethics, 34*, 27). https://doi.org/10.1007/s10806-021-09868-x

Lee, K., & Brummel, Z. L. (2013). Operationalizing the one health approach: The global governance challenges. *Health Policy Plan, 28*(7), 778–785.

Leopold, A. (1949). *A sand county almanac and sketches here and there*. Oxford University Press.

MacPherson, C. (2007). Global bioethics: Did the universal declaration on bioethics and human rights miss the boat? *Journal of Medical Ethics, 33*, 588–590.

Mangena, F. (2013). Discerning moral status in the African environment. *Phronimon, 14*(2), 25–44.

Masaka, D. (2019). Moral status of non-human animals from an African perspective: In defense of moderate anthropocentric thinking. In M. Chemhuru (Ed.), *African environmental ethics* (pp. 223–237). The International Library of Environmental, Agricultural and Food Ethics, Springer.

Potter, V. R. (1971). *Bioethics: Bridge to the future*. Prentice Hall.

Proctor, J. D. (1992). *The owl and the forest*. Ph.D. dissertation, University of California, Berkeley.

Rawlinson, M. C. (2015). Bioethics: A bridge to the future? In G. Solinis (Ed.), *Global bioethics: What for?* (pp. 31–34). United Nations Educational, Scientific and Cultural Organization.

Resnik, D. (2009). Bioethics and global climate change. *Bioethics Forum, 39*(3), 1.

Ritchie, C. (2016). *Moving environmental bioethics into the 21st century: Green bioethics and the common good* [Doctoral Dissertation, Boston College].
Schwabe, C. (1984). *Veterinary medicine and human health*. Williams & Wilkins.
Singer, P. (1990). *Animal liberation*. Random House.
Tosam, M. (2019). African environmental ethics and sustainable development. *Journal of Philosophy, 9*(2), 172–192.
Tsang, Y. (2020, March 16). Hundreds of bats culled in Indonesia to 'prevent spread' of the coronavirus, *South China Morning Post*. https://www.scmp.com/hundredsbats-culled-indonesia-prevent-spread-coronavirus
UNESCO. (2005). *Universal declaration on bioethics and human rights*. UNESCO.
White, L. (1967). The historical roots of our ecologic crisis. *Science, 55*(3767), 1203–1207.
Woodroffe, R., Astrey, K., Barnecut, R., Brotherton, P. N. M., Donnnelly, C. A., Grub, H. M. J., Ham, C., Howe, C., Jones, C., Marriot, C., Miles, V., Rowcliffe, M., Shelley, T., & Truscott, K. (2024). Farmer-led badger vaccination in Cornwall: Epidemiological patterns and perspectives. *People and Nature, 6*(5), 1960–1973. https://doi.org/10.1002/pan3.10691

# Part III
# Reproductive Technologies

# Chapter 6
# Moral Limitations and the Right to Reproduce in the Kenyan Law: Is Kenya Ready for Legalization of Abortion?

**Lillian Omutoko, Melba Katindi, and Walter Jaoko**

**Abstract** Legal access to safe abortion is fundamental for enjoying the rights guaranteed in the Constitution of Kenya, including the rights to life and health. The UN Committee on Economic, Social, and Cultural Rights (CESCR) recognizes reproductive rights as an integral part of the right to health and identifies the denial of abortion as a violation of the right to life. However, the Human Rights Committee's comment on the right to life acknowledges that states, in exercising their sovereignty, can regulate the voluntary termination of pregnancy. Article 26 of the Constitution of Kenya guarantees the right to life. Sub-article 2 states that life begins at conception and outlaws abortion in sub-article 4. Nevertheless, sub-article 4 of the same article contains exceptions and permits abortion under specific conditions, such as when the life and health of the mother, which encompasses physical, psychological, mental, and emotional health, are at risk. Article 26(4) also makes provision to introduce additional circumstances for abortion through legislation. There have been progressive policy measures taken to provide access to safe reproductive health services, including efforts to ensure safe abortion in Kenya. However, attempts at legislation on reproductive rights have been problematic, particularly where provisions include abortion. The 2019 Reproductive Health Bill is the latest of several attempts to address the right to reproductive health care. The objectives of this article are to (1) examine the attempts that have been made to legalize abortion; (2) highlight efforts that have been made to address the right to abortion; and (3) discuss the morality and ethics of abortion in Kenya.

L. Omutoko
Department of Educational Management, Policy and Curriculum Studies, University of Nairobi, Nairobi, Kenya
e-mail: joyce.lillian@uonbi.ac.ke

M. Katindi (✉)
Katindi & Co. Advocates, Nairobi, Kenya
e-mail: mkatindi@katindilawyers.co.ke

W. Jaoko
KAVI-Institute of Clinical Research, University of Nairobi, Nairobi, Kenya
e-mail: wjaoko@kaviuon.org



L. Omutoko, W. Jaoko (eds.), *Bioethics from the Global South*, Advancing Global Bioethics 21, https://doi.org/10.1007/978-3-031-77669-4_6

**Keywords** Abortion · Infertility · Legalization · Morals · Religion · Reproductive health

## 1 Introduction

The debate around abortion remains a contentious issue globally, sparking both criticism and support, and highlighting substantial knowledge gaps and bioethical concerns that warrant a thorough examination. In Western countries, abortion is generally accepted or tolerated within the context of sexual and reproductive health (SRH), based on the principle of autonomy, which upholds a woman's right to self determine whether or not to terminate a pregnancy. This position is championed by the "pro-choice movement," which includes advocates who support abortion with certain caveats, such as limiting the procedure to the first 20 weeks of pregnancy, thus balancing autonomy with the moral status of the fetus (Winter, 2013).

Conversely, a strong antiabortion lobby exists, arguing that abortion equates to taking a life, thus violating the principle of non-maleficence (do no harm) and the right to life of the fetus (Marquis, 2006; Kaczor, 2022). This "pro-life movement" is diverse, encompassing a range of views. Some proponents allow exceptions in cases of rape or incest, where the principle of beneficence (doing good) may outweigh non-maleficence for the pregnant woman (Winter, 2013; Beckwith, 2007). This perspective highlights the belief that while the unborn child's life is sacred, the severe psychological and physical trauma experienced by the woman in such cases justifies an exception. Moreover, the pro-life argument often incorporates the potentiality principle, which suggests that the fetus has the potential to become a fully developed human being and therefore should be afforded moral consideration and rights (Finnis, 2011). This viewpoint aligns with certain interpretations of natural law and religious doctrines, which hold that life begins at conception and must be protected (George & Tollefsen, 2008).

In countries like the United States, confrontations between pro-choice and pro-life groups can be intense and sometimes violent, with abortion being a contentious legal and political issue. The divide has significantly influenced presidential campaigns, distinguishing liberals who are pro-abortion from conservatives who are antiabortion, reflecting the knowledge gaps and lack of consensus on the ethical principles at stake (Student News Daily, 2010). However, the situation in Africa, and specifically Kenya, is dire, as unsafe abortions disproportionately affect the continent, violating the principle of justice by imposing a greater burden on vulnerable populations. Forty-four percent of deaths from unsafe abortions occur in Africa, with 35% of maternal deaths in Kenya linked to unsafe abortions, highlighting the need to address this issue from a public health perspective (Centre for Reproductive Rights, 2010).

Despite these alarming statistics, the debate in Africa is often framed around religion, societal norms, and morality, rather than bioethical principles. For instance, Pope Francis has condemned abortion as a crime (Tollefsen, 2015; Vatican News, 2021), reflecting a view shared across various faiths and by some secular individuals, prioritizing the principle of non-maleficence over autonomy. On the other hand, organizations like Marie Stopes argue that abortion is a critical aspect of comprehensive health care, emphasizing the principle of beneficence and the right to access safe reproductive services. In Kenya, Marie Stopes Clinics have provided contraception and abortion services since 1985, but their activities were banned in 2018 by the Medical Practitioners and Dentists Board, citing the promotion of abortion in a country where it is illegal unless the mother's health is at risk. The ban raised concerns about the balance between autonomy and non-maleficence (BBC News, 2018).

This ban also raises the question of whether Kenya is prepared for the legalization of abortion, and whether such a move would uphold the principles of bioethics. The ethical debate centers on whether legalizing abortion addresses health concerns arising from unsafe procedures, promoting beneficence, or if it merely seeks to control reproductive rights, which could be seen as restricting autonomy. Proponents of legalizing abortion argue that criminalizing it drives women to seek unsafe procedures, leading to complications and deaths, thus violating the principles of non-maleficence and justice (Grimes et al., 2006; Loi et al., 2015). For instance, the Kenya Demographic and Health Survey (2008–2009) reported about 488 maternal deaths per 100,000 live births, with unsafe abortions being a significant contributor, highlighting the need for interventions that uphold the principles of beneficence and justice (Kenya National Bureau of Statistics and ICF Macro, 2010). Furthermore, controlling reproductive rights through criminalizing abortion could be perceived as a restriction on women's autonomy and their ability to make decisions about their own bodies and reproductive choices, rather than prioritizing autonomy.

While groups advocating for “safe abortion” present data to support their claims, these figures are often disputed by pro-life supporters, who argue they are exaggerated to push for legalization, reflecting the knowledge gaps and lack of consensus on the ethical principles at stake. Estimates indicate over 4 million unsafe abortions occur annually in Africa, with the continent having the highest risk of maternal death from it, raising concerns about justice and beneficence (Gebremedhin et al., 2018). These discrepancies highlight the polarized opinions and underscore the need for a deeper understanding of the ethical and moral implications, as well as the application of bioethical principles in the context of abortion in Kenya.

The critical ethical questions include the moral justification for a broad interpretation of abortion rights in Kenya, balancing the autonomy of pregnant women versus the right to life of the unborn child, and the potential obligation of healthcare providers to uphold the principles of beneficence and non-maleficence. Additionally, concerns about the impact of legalizing abortion on societal values, such as the potential for gender selection and other moral issues, need addressing from the perspective of justice and beneficence. The African context also emphasizes reproduction as a cultural obligation, complicating the abortion debate with deeply rooted societal norms that may conflict with the principles of autonomy and beneficence.

This chapter examines the abortion debate through the lens of Kenya, exploring regional perspectives, historical attempts at legalization, and current parliamentary discussions. It aims to highlight the moral and ethical issues that arise within Kenyan society whenever this contentious topic is deliberated. To achieve the above objectives, this chapter employed a comprehensive literature review, drawing from academic databases such as PubMed, Embase, and Google Scholar, as well as reports from international organizations. The review focused on global bioethics literature, legal and policy documents, and empirical studies related to the abortion debate, particularly in the African context. The authors sought to synthesize the existing knowledge, identify key gaps, and propose a global bioethics framework for further analysis of this critical public health and human rights issue.

## 2 Abortion Law in Kenya

This section evaluates the international laws and national legal frameworks crucial for determining the context within which abortion is acceptable. Tracing the trajectory of abortion legalization from the 2010 Constitution and the Reproductive Health Bill highlights significant knowledge gaps and bioethical issues. Although the Bill has not been enacted, its discussion underscores the need to address unintended pregnancies and unsafe abortions, impacting attitudes and practices of individuals, professionals, and institutions. Kenya recognizes ratified international treaties and conventions guiding legal and moral decisions. This provides a foundation for the state to affirm the provision of safe abortion as a fundamental human right (Fine et al., 2017). While the Convention on the Elimination of All Forms of Discrimination against Women mandates states to provide family planning and prohibits discrimination against women, thereby promoting justice in reproductive health services (United Nations, 2014b), in Kenya, the lack of clear legal guidelines creates inequalities in access to safe abortion, disproportionately affecting marginalized and vulnerable groups, which contrasts with the principles of justice and non-discrimination promoted by the Convention (Centre for Reproductive Rights, 2010).

Regionally, the Protocol to the African Charter on Human and Peoples' Rights on the Rights of Women in Africa, also known as the Maputo Protocol, makes provisions requiring protection of the right to abortion "in cases of sexual assault, rape, incest, and where the continued pregnancy endangers the mental and physical health of the mother or the life of the mother or the fetus." However, Kenya signed and ratified the protocol with reservations pertaining to the medical termination of pregnancy on the basis that it expands abortion beyond the strict limitations in the Constitution, which remains the supreme law. Other than the Maputo Protocol, a right to safe abortion is not explicitly recognized in any international treaty. In light of this, United Nations Treaty bodies and agencies have inferred the right to safe abortion services from other protected human rights: mainly the rights to life and health, but also other rights such as freedom from discrimination and the right to privacy, contained in international law (Stevens, 2018).

The International Covenant on Economic, Social and Cultural Rights guarantees the right to the highest attainable standard of health, which is understood to include the freedom to make choices about one's health. The Convention on the Elimination of All Forms of Discrimination against Women requires states to provide family planning within healthcare services. Additionally, the Convention specifically prohibits discrimination against women (United Nations, 2014b). The human rights committee's general comment on the right to life requires states to adopt measures to provide safe abortion, and failure to do so is considered a violation of the right to life (Joseph, 2019). It is argued that the push for a legal framework on abortion aims at promoting human rights of women and girls by addressing problems related to the termination of pregnancy. It is also argued that legalization can reduce the number of women who die every year due to unsafely performed induced abortions (Obengo, 2016).

The legal framework in Kenya is similar to many African countries that retained the punitive measures in penal codes inherited from English common law during the colonial period (Berer, 2017). Most countries in the region either completely criminalize abortion or permit abortion under restrictive circumstances to preserve the life of a pregnant woman. This is consistent with the widely accepted decision in the case 1938 R v. Bourne of London, which informed the application of abortion provisions in criminal law in African Commonwealth member states (Cook & Dickens, 1981). Countries that have adopted new constitutions or reviewed their laws more recently have included a national standpoint on abortion allowing it on grounds relating to safeguarding the life and health of the mother. Still, few of these countries have actually developed specific legislation to effect and implement accessible safe termination of pregnancy, and only a handful of countries in Africa liberally permit abortion without imposing any legal restrictions.

World Health Organization's (WHO) guidance on abortion is informed by evidence that fewer deaths occur from unsafe abortions in countries whose legal framework contains broad provisions allowing for abortion (WHO, 2012). Hence, where the Constitution provides a framework for abortion, the development of specific abortion laws is critical in interpreting and applying justifications for the termination of pregnancy beyond the narrow focus on saving the life of the mother. The scope of abortion law is critical in determining how the right to safe abortion is availed; it may broaden permissible grounds to include mental health and social circumstances. Additionally, legislation would regulate and standardize the provision of safe abortion services by addressing related technical, structural, and procedural problems that would limit wider access.

With increasing awareness of the impact of a punitive legal environment on unsafe abortion, African countries have progressively reformed their laws to not only allow but also expand the grounds for abortion beyond saving the life and physical health of a woman to include instances of rape, incest, and serious fetal conditions, as is the case in Ethiopia and Zambia (Blystad et al., 2019). The liberalization of abortion law in Ethiopia amid religious antiabortion viewpoints has resulted in a significant improvement in the delivery of abortion services (Gebrehiwot et al., 2016). Similarly, despite cultural positions against abortion, Zambia is one of the countries that has expanded the circumstances under which abortion is legally permissible to include broad health and socio-economic grounds (Blystad et al., 2019). Tunisia, which was the first Muslim country to permit abortion for women in certain circumstances, further liberalized its law to allow abortion for all women in the first trimester (Hajri et al., 2015). South Africa is among a handful of countries in the region that have adopted the most liberal legal approach to abortion. Termination of pregnancy by women in South Africa is permitted without question in the first trimester and on very expansive grounds in later stages of pregnancy, including medical, economic, and social reasons. A Christian lawyers group lost two cases that sought to challenge the constitutionality of the law, facilitating the expansion of qualified medical staff who are legally allowed to perform abortions in early pregnancy to include certified midwives and registered nurses in primary health care facilities (Albertyn, 2015). Interestingly, the law in South Africa criminalizes any interference with or access to legal abortion.

On the contrary, the absence of a specific enabling abortion law creates uncertainty around the circumstances in which termination of pregnancy is permissible in Kenya. Whereas the prohibition of abortion in the Constitution of Kenya, 2010 contains exceptions around the necessity to save the life and health of the mother, it also anticipates other grounds as provided in law. Even though the Health Act makes mention of abortion, no supportive provisions to inform "other" circumstances and the delivery of abortion services are contained in law. With the Penal Code containing provisions criminalizing actions performed "unlawfully" by any person in the termination of pregnancy (Kenya Law, 2014), there is a need for clarity on what comprises "lawful abortion." This has resulted in the arrest and prosecution of women seeking and health providers offering abortion services.

Perceptions of abortion as punishable by law continue to negatively affect social decisions in the demand and provision of abortion services, thereby restricting required medical services as has previously been documented (Centre for Reproductive Rights, 2010). Data collected by the Centre for Reproductive Rights Kenya indicate that there were about three cases every week in which women were charged for having illegal abortions. The cases involved girls and women, some of whom were minors. Most of the women who were arrested for abortion-related crimes were given probation rather than imprisonment, and those who could not afford bail remained in remand for about a year. In 2013, a health professional and another person were charged with attempting to procure an abortion contrary to the Penal Code, but the case was dismissed in 2016 because there was neither a police file nor a witness (International Campaign for Women's Right to Safe Abortion, 2017).

Prior to the Constitution of Kenya, 2010, there were several attempts at achieving legislative clarity on abortion and its permissibility in Kenya. This push was predominantly led by women's rights organizations and health service providers whose arguments are based on saving lives and preventing harm to the mother and to society (Wambui, 2016). Some researchers have documented the negative impact of unsafe abortion practices, and these have formed the basis of calls demanding for the availability and standardization of safe abortion services (Mumah et al., 2014; Mutua et al., 2018). However, opponents of abortion in the country have been from many quarters, including physicians, legal experts, politicians, philosophers, and other healthcare professionals. One of the most vocal opposition groups has been religious leaders who have continuously challenged pro-choice arguments on the basis of the sanctity of life and the need to safeguard the fetus (Obengo, 2016).

In 2004, religious leaders objected to an attempt to include reproductive health rights as well as abortion in the draft Constitution, and the draft ultimately failed. An attempt to revive provisions to permit abortion in certain circumstances in the Reproductive Health and Rights Bill of 2008 was unsuccessful. There has been quite some resistance to the legalization of abortion, and in 2008, a pastoral letter was issued by the Archbishop of the Catholic Archdiocese of Nairobi warning the Government against passing the Bill. The letter stated in part that abortion is not merely the removal of some tissue from a woman's body but the removal of a living thing that would become human if it were allowed to remain inside the woman's body, that it is the destruction of an unborn baby, and that pregnancy is the period for this new human life to mature, not just to become human. In other words, the fetus already is human. These statements emphasize the perspective of religious leaders, which resonates with pro-life ethicists who argue that abortion is morally wrong. The ethical concern from a religious creationist perspective is who creates life and who has a right over another's life, as well as whether a mother has a right to decide to keep or not to keep a pregnancy. These questions raise health concerns about the effects of unsafe abortions (Obengo, 2016).

In 2009, negotiations on a new Constitution resumed after the 2004 draft was rejected in a referendum. During this time, religious leaders and antiabortion supporters insisted on the inclusion of a clause specifying that life begins at conception. The most contested provision by religious leaders was that the Constitution ought to be clear on the protection of the sanctity of human life and also state where life begins and where it ends. Otherwise, it risks portraying Kenya as an immoral state, a position that would be contested by pro-choice ethicists (Mathenge, 2009). The need for clarity resulted in a clause that recognizes life as beginning at conception and a declaration that abortion is illegal. However, a compromise clause was also included to narrowly qualify circumstances where abortion would be acceptable.

After all the deliberations and attempts to stop the legalization of abortion, The Constitution of Kenya, 2010 permits abortion in limited circumstances. Article 26(4) of the Constitution, while generally prohibiting abortion, expanded the scope of lawful abortion, permitting a trained health professional to determine necessity for emergency treatment, or to prevent loss of life or harm to the mother's health. The Constitution also contemplates the enactment of specific legislation that could provide other permissible grounds in which abortion would be lawful.

In 2014, Parliament attempted to pass the Reproductive Health Care Bill, but it was unsuccessful. The bill was criticized for not adequately defining the legal limits of abortion as outlined in Article 26(4) of the Constitution. Instead, it merely reiterated the constitutional provisions on terminating pregnancy without addressing other related aspects. Currently, Kenya is making another attempt to pass a comprehensive law on reproductive rights, including provisions on abortion. The Reproductive Health Bill of 2019 aims to realize the right to reproductive health as provided in Article 43, which is a component of the right to health in Kenya (Ajayi & Muoka, 2020). This bill grants all individuals, including adolescents, the right to make decisions about their reproductive health. Part IV of the bill specifically addresses the termination of pregnancy. In addition to allowing abortion for emergency treatment and to protect the life and health of the mother, as stipulated in the Constitution, the bill also permits abortion in cases of fetal abnormalities. However, there are fears that including provisions on abortion within the context of the right to reproductive health guaranteed in Article 43 poses a risk of expanding its legal limits beyond the constitutional threshold. The inclusion of fetal abnormality as a ground for abortion and the reproductive rights of adolescents has once again divided the country, with religious and cultural objections arising. Various parties, including parliamentarians and religious leaders, have attacked the bill for attempting to expand the scope of abortion grounds beyond what is outlined in the Constitution and in opposition to country-specific religious and cultural values.

Objections to efforts aimed at providing legal interpretations of permissible grounds for abortion in Kenya have hindered legislative progress. In the absence of clear laws, the High Court has been relied upon to interpret abortion rights. The High Court, by enforcing the progressive human rights provisions in the Constitution's Bill of Rights, has played a crucial role in advancing the legal discourse on reproductive rights for women and girls. The most significant ruling by the court occurred on June 12, 2019, in the case Federation of Women Lawyers (FIDA – Kenya) & 3 others v. Attorney General & 2 others; East Africa Center for Law & Justice & 6 others (Interested Party) & Women's Link Worldwide & 2 others (Kenya Law, 2019). Women's rights organizations protested the death of a teenager following a failed abortion procedure. The girl had been forced into sexual intercourse by an older man, resulting in her pregnancy. The case was filed to challenge the Kenyan government's withdrawal of the "Standards and Guidelines for Reducing Morbidity and Mortality from Unsafe Abortion in Kenya" (Ministry of Medical Services, 2012), claiming that this action undermined access to safe legal abortion services. The government had withdrawn the guidelines shortly after their publication due to disputes among stakeholders regarding their content. In this case, the court ruled that abortion is permissible for victims of sexual violence.

The legislative debate finds a convergence on restricting the availability of abortion as a medical necessity to save the life or health of the pregnant woman and upon the determination of a qualified health care worker or in pregnancy resulting from rape or defilement. The divergence is premised on the possibility of legislating on broader grounds which would provide other medical and social circumstances for which a pregnant woman or girl could terminate a pregnancy. Ultimately, the discourse around whether or not abortion is permissible as a moral choice will largely inform Kenya's legislative approach on whether or not to broaden the permissible grounds for the termination of pregnancy. The next section discusses the ethical and moral issues from human rights and religious and cultural perspectives.

## 3 Should Kenya Legalize Abortion? The Moral Question

Unsafe abortion is a global public health problem, and denying or neglecting it does not reduce or eradicate the ethical dilemmas surrounding abortion (Fadhalla, 2019). Abortion remains a controversial topic in reproductive health (Cook & Dickens, 2003). Annually, 7% per 1000 women in developing regions are treated for complications arising from induced abortion. In 2012, 48% per 1000 women in Kenya had unsafe abortions. Women risk dying from untreated complications when they do not get the care they need, with the highest case-fatality rates being in Africa at 141 per 1000 (WHO, 2008). Unintended pregnancies in Kenya are at 120 per 1000, many of which likely end in abortion, potentially leading to death without accessible post-abortion care (Mohamed et al., 2015). In 2015, the unmet need for modern contraception in Africa was 26% (Singh et al., 2017). Unsafe abortion is a leading cause of maternal death in Kenya, with a maternal mortality ratio of 362 per 100,000 among women aged 25–39, and 17% of these deaths are associated with induced abortion (Kenya National Bureau of Statistics, 2014; Say et al., 2014; Mutua et al., 2018). This scenario raises the question of whether Kenya should legalize safe abortion. The discussion largely revolves around the morality of abortion in relation to women's rights, human rights, religion, and culture.

Achieving readiness for the legalization of abortion in Kenya necessitates widespread societal acceptance, where the majority of the population supports abortion laws and can access safe services without fear or stigma. However, this readiness remains elusive due to ongoing debates and unmet family planning needs, hindering the establishment of a conducive environment for safe abortion access. Even in contexts where abortion has been legalized, ensuring the sustainability of safe services requires well-equipped healthcare systems and trained staff to safeguard women's health. Unfortunately, several third world countries lack the necessary resources to sustain such infrastructure and support, presenting a significant barrier to achieving the intended purpose and raising the question of equitable access.

While international human rights law does not specifically recognize abortion as a right, arguments in favor of abortion are inferred based on other recognized rights. International human rights instruments protect and promote reproductive rights without discrimination. They recognize the right of all to decide freely and responsibly the numbers, spacing, and timing of their children and related matters. Although these rights are well recognized, the United Nations Population Fund (UNFPA) aims to increase access to sexual and reproductive health, secure reproductive rights, and reduce maternal mortality, with the ultimate goal of improving the lives of adolescents, youth, and women. The UNFPA aims to deliver a world where every pregnancy is wanted, every birth is safe, and every young person's potential is fulfilled without explicitly supporting the termination of pregnancies (United Nations, 2014a). The United Nations Committee on the Elimination of Discrimination against Women argues that women alone have the right to decide whether to have an abortion or not (United Nations, 1999). Thus, it can be argued that a pregnant woman has the right to independently make decisions, and a violation of these rights poses a threat to several human rights including the right to health, equality, liberty, non-discrimination, and privacy (WHO, 2017; United Nations, 2014a, b). Additionally, advocates for abortion rights argue that women should have the autonomy to make choices about their reproductive health without interference from the state or other external actors. However, autonomy in reproductive choices, including abortion, remains a challenge in Kenya. Cultural, religious, and legal barriers limit women's ability to make independent decisions about their reproductive health, posing a threat to their rights to health, equality, and privacy. Legalizing abortion could empower women to exercise their autonomy and access safe healthcare services, but entrenched societal norms and inadequate legal frameworks hinder progress in this area.

Reproductive health is a fundamental human right inherent to all individuals, as outlined in the Universal Declaration of Human Rights (UDHR). Although not legally binding, the UDHR establishes basic obligations for states and citizens, emphasizing the equal dignity and rights of all members of the human family, including women and children (United Nations, 2005; Deaton, 2011). However, in Kenya, this right must be balanced with the constitutional declaration that life begins at conception, intertwining the rights of women with those of the unborn child (Constitution of Kenya, 2010). Despite this, the current legal framework in the country limits access to abortion services, and it has been argued that this results in unjust outcomes for women in need. The alarming rates of maternal mortality due to unsafe abortions highlight the pressing need for a more equitable approach to reproductive healthcare in the country (United Nations, 2005; Deaton, 2011).

Abortion is a dreaded secret in several African countries, often relegated to the darkest corners of fear and mythology. For some societies, it remains untouched and undiscussed. Since the 1970s, legislation has specified when abortion can be lawfully conducted in developed and developing countries. The indications for legal abortion have been broadly divided into medical, eugenic, humanitarian (e.g., where the pregnancy is a product of a criminal act such as rape or incest), and medico-social indications. In Kenya, considerations for abortion are mainly medical, with arguments for expanding the scope of legalization to include pregnancies resulting from rape or incest. Fears have been expressed regarding the lack of clarity, which could open room for sex selection and prenatal implantation diagnosis where embryos are genetically profiled before implantation in assisted reproduction technology. Within the Kenyan context, these fears may be propagated by multiple cultures and religions that hold a common view on the sacredness of life and the right to life, a situation that may be contested by couples experiencing infertility and desiring healthy children (WHO, 1971).

The pertinent questions involve when personhood begins: is it after birth, at birth, during pregnancy, or at conception? These concerns are raised by religious groups when addressing the beginning of life. When it comes to human rights, questions include whether the human embryo has rights, whether it is wrong to abort a person, and whether abortion is permissible in hard cases. Hard cases for critics of abortion include fetal deformity, abortion for the child's good, cases of rape and incest, and abortion to save the mother's life (May, 2012). With globalization and increased interactions, cultures have become fluid with varying belief systems emerging. The big question is whether abortion is permissible and if so, when? Hard cases for defenders of abortion include sex-selection abortion and abortions due to unmet family planning needs, which is a major cause of abortions among teenagers and young adults. These circumstances raise debates on whether abortion is morally permissible or objectionable (May, 2012). In the attempt to legalize abortion, Kenya is caught up in debates marred by differing perspectives.

Kenya, like several other African states, has retained abortion legislation introduced by colonial countries, authorizing abortion only when necessary to preserve the life of the pregnant woman ( Constitution of Kenya, 2010; WHO, 1971). However, Kenya has unsuccessfully made several attempts to legalize abortion in view of women's rights, the right to life, and justice. The question of readiness for the legalization of abortion in Kenya brings about the morality argument. Legalization may not be the answer to unsafe abortions. Some developing countries have legalized abortion, but inadequate health systems fail to meet the demand, leading women to seek services from unqualified doctors or inappropriate facilities. Such situations expose women to harm, violating the principle of non-maleficence. Where abortion has been legalized, health systems need to be well equipped and staffed to protect women seeking abortion services from harm. From a public health perspective, there is an ethical imperative to protect women's rights and their reproductive, emotional, and mental health. Decisions around the morality and orality of abortion should be based on the magnitude of health problems caused by unsafe abortions, health inequities, and social injustices women face when they cannot access safe abortion. The voices of philosophers, religious groups, lawyers, government, and the public lean on their respective convictions, encapsulated in belief systems, religion, and culture. The debate on the morality of the legalization of abortion must consider how abortion intersects with morality, religion, law, human life, human rights, and justice (Fadhalla, 2019; May, 2012).

The context of legalization of abortion requires reflection on law, conventions, social and religious norms, family contexts, and relationships. The ethics of abortion should form significant grains of whether the legalization of abortion is moral or not, although not everyone will agree on it. Feminists argue from a liberal rights perspective based on individual autonomy, asserting that women have the right to decide what happens to their bodies. Proponents of abortion base their arguments on the principles of equality, rights, and freedom of choice. Human rights encompass reproductive freedoms for all women, including whether to keep a pregnancy or not (Porter, 1994). Furthermore, the debate on the legalization of abortion needs to appeal to reason rather than solely religious beliefs, especially where there are high maternal and child mortality rates and other health risks such as low rates of access to contraception (Latt et al., 2019; Feyssa & Gebru, 2022; World Health Organization, 2024).

The vulnerability of women due to gendered power dynamics raises significant ethical concerns in the abortion debate. These power imbalances often lead to unprotected sexual intercourse and unintended pregnancies, particularly when family planning needs are unmet. Consequently, the increased demand for abortion arises, yet safe abortion services are often inaccessible due to restrictive laws or prohibitive costs (Fadhalla, 2019). This situation highlights issues of justice, as women are disproportionately affected by barriers to reproductive healthcare. Moreover, the historical context of abortion legislation, compounded by diverse cultural, religious, legal, and belief systems, perpetuates the ongoing debate. In navigating these complexities, the focus should prioritize saving the lives of women and girls from the dire consequences of unsafe abortions, which pose significant risks to their health and burden the healthcare system with high costs associated with managing complications.

Discussions surrounding the morality of legalizing abortion in Kenya within the framework of human rights encounter a complex interplay of ethical principles. The question of whose rights—those of the woman or the unborn child—take precedence presents a moral dilemma. From a deontological perspective, which emphasizes duties and obligations, some argue for prioritizing the protection of the vulnerable party, the fetus, particularly in situations of conflict between rights. This perspective emphasizes the inherent value and rights of the fetus as a potential human life. Conversely, a utilitarian approach weighs the consequences of actions and aims to minimize harm. Here, the focus shifts to assessing which decision—protecting the woman's rights or preserving the fetus—leads to the least overall harm. Additionally, considerations of justice come into play, as different stakeholders advocate for their respective rights. Religious beliefs, particularly within Christian groups in Kenya, often oppose abortion on moral grounds, equating it with murder and emphasizing the sanctity of life from conception (May, 2012). However, feminist perspectives argue for the autonomy of women to make decisions about their own bodies, prioritizing women's rights and bodily autonomy. They contend that abortion is morally justifiable, especially in cases where it is necessary to safeguard women's lives or enable them to control their reproductive choices (Fadhalla, 2019). Despite these differing viewpoints, the question of the moral acceptability of abortion remains unresolved, underscoring the complexity of ethical considerations in this debate.

Debates surrounding the morality of legalizing abortion in Kenya are multifaceted, influenced by a confluence of human rights, religious, cultural, and belief perspectives. From a bioethical standpoint, the issue revolves around principles of justice and non-maleficence. Despite the recognition of reproductive rights in international human rights instruments, Kenya grapples with conflicting interpretations of whose rights should be prioritized. The lack of consensus exacerbates the challenge, especially concerning women's rights to access safe abortion and the protection of the unborn child's rights. Additionally, the inadequacy of Kenya's healthcare system to meet the demand for safe abortion services underscores concerns related to non-maleficence, as evidenced by the high incidence of unsafe abortions and associated maternal and child mortality rates. For example, in 2015–2019, there were a total of 2,380,000 pregnancies annually. Of these, 1,450,000 pregnancies were unintended and 551,000 ended in abortion (Guttmacher Institute, 2022). While the rate of abortion continues to rise, the discordance between the principles of justice and non-maleficence further complicates the path toward legalization, reflecting the ethical complexities inherent in this contentious issue.

## 4 Conclusion

A lot of debates in literature on abortion law and practice are largely from the Global North. However, these discussions are becoming louder in the Global South (Erdman, 2017). In the Kenyan context these debates have largely been in the context of the proposed Reproductive Health Bill currently in discussion. From a moral perspective, considering rights, religion, and culture in Kenya, the debates on abortion over several years indicate that the country is not ready for the legalization of abortion despite the importance of women's reproductive health rights. To mitigate against the negative social impact of unwanted pregnancies, we propose strengthening of sex education to prevent the pregnancies and adoption care for the children born in circumstances where they are unwanted.

## References

Ajayi, A., & Mwoka, M. (2020). Academic rigour journalistic flair, Kenya is having another go at passing a reproductive rights bill. *The Conversation*. https://theconversation.com/kenya-is-having-another-go-at-passing-a-reproductive-rights-bill-whats-at-stake-142387. Accessed 10 Sept 2020.

Albertyn, C. (2015). Claiming and defending abortion rights in South Africa. *Revista Direito GV, 11*(2), 429–454.

BBC News. (2018). *Kenya bans Marie Stopes from offering abortion services*. https://www.bbc.com/news/world-africa-46254630

Beckwith, F. J. (2007). *Defending life: A moral and legal case against abortion choice*. Cambridge University Press.

Berer, M. (2017). Abortion law and policy around the world: In search of decriminalization. *Health and Human Rights Journal, 19*(1), 13–27.

Blystad, A., Haukanes, H., Tadele, G., Haaland, M. E., Sambaiga, R., Zulu, J. M., & Moland, K. M. (2019). The access paradox: Abortion law, policy and practice in Ethiopia, Tanzania and Zambia. *International Journal for Equity in Health, 18*(1), 126.

Centre for Reproductive Rights. (2010). *In harm's way: The impact of Kenya's restrictive abortion law*. https://reproductiverights.org/in-harms-way-the-impact-of-kenyas-restrictive-abortion-law/

Constitution of Kenya. (2010). http://kenyalaw.org:8181/exist/kenyalex/actview.xql?actid=Const2010. Accessed 12 Sept 2020.

Cook, R. J., & Dickens, B. M. (1981). Abortion laws in African Commonwealth countries. *Journal of African Law, 25*(2), 60–79.

Cook, R. J., & Dickens, B. M. (2003). Human rights dynamics of abortion law reform. *Human Rights Quarterly, 25*, 1–59.

Deaton, M. (2011). Universal declaration of human rights. In D. K. Chartejee (Ed.), *Encyclopaedia of global justice*. Springer.

Erdman, J. (2017). Theorizing in abortion law and human rights. *Health and Human Rights Journal, 19*(1), 29–40.

Fathalla, M. F. (2019). Abortion and public health ethics. In *The Oxford handbook of public health ethics*. https://doi.org/10.1093/oxfordhb/9780190245191.013.34. https://www.oxfordhandbooks.com/view/10.1093/oxfordhb/9780190245191.001.0001/oxfordhb-9780190245191-e-34? print=pdf

Feyssa, M. D., & Gebru, S. K. (2022). Liberalizing abortion to reduce maternal mortality: Expanding access to all Ethiopians. *Reproductive Health, 19*(S1). https://doi.org/10.1186/s12978-022-01457-z

Fine, J. B., Mayall, K., & Sepúlveda, L. (2017). The role of international human rights norms in the liberalization of abortion laws globally. *Health and Human Rights, 19*(1), 69–80.

Finnis, J. (2011). *Natural law and natural rights*. Oxford University Press.

Gebrehiwot, Y., Fetters, T., Gebreselassie, H., Moore, A., Hailemariam, M., Dibaba, Y., & Getachew, Y. (2016). Changes in morbidity and abortion care in Ethiopia after legal reform: National results from 2008 and 2014. *International Perspectives on Sexual and Reproductive Health, 42*(3), 121–130.

Gebremedhin, M., Semahegn, A., Usmael, T., & Tesfaye, G. (2018). Unsafe abortion and associated factors among reproductive aged women in Sub-Saharan Africa: A protocol for a systematic review and meta-analysis. *Systematic Reviews, 7*(1), 130.

George, R. P., & Tollefsen, C. (2008). *Embryo: A defense of human life*.

Grimes, D. A., Benson, J., Singh, S., Romero, M., Ganatra, B., Okonofua, F. E., et al. (2006). Unsafe abortion: The preventable pandemic. *The Lancet, 368*(9550), 1908–1919.

Guttmacher Institute. (2022). *Kenya country profile*. Guttmacher Institute. https://www.guttmacher.org/regions/africa/kenya

Hajri, S., Raifman, S., Gerdts, C., Baum, S., & Foster, D. G. (2015). 'This is real misery': Experiences of women denied legal abortion in Tunisia. *PLoS One, 10*(12), e0145338. https://www.safeabortionwomensright.org/wpcontent/uploads/2018/02/Standards-Guidelines-for-the-Reduction-of-Morbidity-and-Mortality-from-Unsafe-Abortion.pdf. Accessed 13 Sept 2020.

International Campaign for Women's Right to Safe Abortion. (2017). *The law, trials and imprisonment for abortion in Kenya*. https://www.safeabortionwomensright.org/news/the-law-trials-and-imprisonment-for-abortion-in-kenya/

Joseph, S. (2019). General Comment No. 36 (2018) on Article 6 of the International Covenant on Civil and Political Rights, on the Right to Life (HR Comm.). *International Legal Materials, 58*(4), 849–871.

Kaczor, C. (2022). The ethics of abortion. *Taylorfrancis*. https://doi.org/10.4324/9781003305217

Kenya Law. (2014). *Kenya Penal Code*. http://kenyalaw.org:8181/exist/kenyalex/actview.xql?actid=CAP.%2063. Accessed 14 Sept 2020.

Kenya Law. (2019). Petition 266 of 2015. Federation of Women Lawyers (Fida – Kenya) & 3 others v Attorney General & 2 others; East Africa Center for Law & Justice & 6 others (Interested Party) & Women's Link Worldwide & 2 others (Amicus Curiae) [2019] eKLR. http://kenyalaw.org/caselaw/cases/view/175490/. Accessed 12 Sept 2020.

Kenya National Bureau of Statistics. (2014). *Kenya demographic and health survey*. Kenya National Bureau of Statistics (KNBS), ICF International.

Kenya National Bureau of Statistics (KNBS) and ICF Macro. (2010). *Kenya demographic and health survey 2008-09*. Calverton.

Latt, S. M., Milner, A., & Kavanagh, A. (2019). Abortion laws reform may reduce maternal mortality: an ecological study in 162 countries. *BMC Women's Health, 19*(1). https://doi.org/10.1186/s12905-018-0705-y

Loi, U. R., Gemzell-Danielsson, K., Faxelid, E., & Klinberg, M. (2015). Health care providers' perceptions of and attitudes towards induced abortions in sub-Saharan Africa and Southeast Asia: A systematic literature review of qualitative and quantitative data. *BMC Public Health, 15*, 139. https://doi.org/10.1186/s12889-015-1502-2

Marquis, D. (2006). Why abortion is immoral. *Bioethics: An Anthology, 2*, 51.

Mathenge, G. (2009). Church faults draft over right to life. *The Standard. Nairobi*, November 22. https://www.standardmedia.co.ke/busia/article/1144028797/church-faults-draft-over-right-to-life. Accessed 11 Jan 2021.

May, W. E. (2012). Book review. The ethics of abortion: Women's rights, human life, and the question of justice. *Linacre Quarterly, 79*(1), 104–117.

Ministry of Medical Services. (2012). *Standards guidelines for reducing morbidity and mortality from unsafe abortion in Kenya*, September 2012.

Mohamed, S. F., Izugbara, C., Moore, A. M., Mutua, M., Kimani-Murage, E. W., Ziraba, A. K., Bankole, A., Singh, S. D., & Egesa, C. (2015). The estimated incidence of induced abortion in Kenya: A cross-sectional study. *BMC Pregnancy and Childbirth, 15*(1). https://doi.org/10.1186/s12884-015-0621-1

Mumah, J., Kabiru, C. W., Mukiira, C., Brinton, J., Mutua, M., Izugbara, C. O. et al. (2014). *Unintended pregnancies in Kenya: A country profile*. https://knowledgecommons.popcouncil.org/departments_sbsr-rh/271/. Accessed 12 Jan 2021.

Mutua, M. M., Manderson, L., Musenge, E., & Achia, T. (2018). Policy, law and post-abortion care services in Kenya. *PLoS One, 13*(9), e0204240. https://doi.org/10.1371/journal.pone.0204240

Obengo, T. J. (2016). *The quest for human dignity in the ethics of pregnancy termination*. Wipf & Stock.

Porter, E. (1994). *Abortion ethics: Rights and responsibilities*. https://www.researchgate.net/publication/11690936_Abortion_Ethics_Rights_and_Responsibilities. Accessed 6 Apr 2021.

Say, L., Chou, D., Gemmill, A., Tunçalp, Ö., Moller, A.-B., Daniels, J., et al. (2014). Global causes of maternal death: A WHO systematic analysis. *Lancet Global Health. [Internet], 2*, e323–e333. https://doi.org/10.1016/S2214-109X(14)70227-X

Singh, S., Remez, L., Sedgh, G., Kwok, L., & Onda, T. (2017). *Abortion worldwide 2017. uneven progress and unequal access*. Guttmacher Institute. https://www.guttmacher.org/sites/default/files/report_pdf/abortion-worldwide-2017

Stevens, A. (2018). Pushing a right to abortion through the back door: The need for integrity in the UN Treaty Monitoring System, and perhaps a Treaty amendment. *Penn State Journal of Law & International Affairs, 6*, 70–141.

Student News Daily. (2010). https://www.studentnewsdaily.com/conservative-vs-liberal-beliefs/. Accessed 30 Aug 2020.

Tollefsen, C. (2015). Pope Francis and abortion. *Christian Bioethics, 21*(1), 56–68. https://doi.org/10.1093/cb/cbu044

United Nations. (1999). Committee on the Elimination of Discrimination against Women (CEDAW): General Recommendation 24: Article 12 of the Convention (women and health) (20th Sess., 1999). http://www.un.org/womenwatch/daw/cedaw/recommendations/recomm.htm#recom24. Accessed 12 Jan 2021.

United Nations. (2005). *The United Nations Human Rights Treaty System. An introduction to the core human rights treaties and the treaty bodies*. https://www.ohchr.org/documents/publications/factsheet30rev1.pdf. Accessed 12 Jan 2021.

United Nations. (2014a). *Reproductive rights are human rights. A handbook for National Human Rights Institutions*. https://www.ohchr.org/Documents/Publications/NHRIHandbook.pdf. Accessed 12 Jan 2021.

United Nations. (2014b). *Women's rights are human rights*. https://www.ohchr.org/Documents/Events/WHRD/WomenRightsAreHR.pdf. Accessed 12 Jan 2021.

Vatican News. (2021, September 15). *Pope: Abortion is murder, the Church must be close and compassionate, not political – Vatican News*. www.vaticannews.va. https://www.vaticannews.va/en/pope/news/2021-09/pope-abortion-is-murder-the-church-must-be-compassionate.html

Wambui, J. (2016). *The pot bearer understands its weight: An analysis of pro-abortion discourses in Kenya*. http://journals.uonbi.ac.ke/index.php/aws/article/view/1529. Accessed 12 Jan 2021.

Winter, A. (2013). Anti-abortion extremism and violence in the United States. In G. Michael (Ed.), *Extremism in America*. University Press Florida.

World Health Organization. (1971). *Abortion laws*. https://apps.who.int/iris/bitstream/handle/10665/41924/a53898_eng.pdf?sequence=1&isAllowed=y

World Health Organization. (2008). *Unsafe abortion: Global and regional estimates of the incidence of unsafe abortion and associated mortality*. Geneva https://www.who.int/reproductivehealth/publications/unsafe_abortion/9789241501118/en/. Accessed 12 Sept 2020.

World Health Organization. (2012). *Safe abortion: Technical and policy guidance for health systems*. World Health Organization. https://www.who.int/reproductivehealth/publications/unsafe_abortion/9789241548434/en/. Accessed 12 Jan 2021.

World Health Organization. (2017). *Human rights and health*. https://www.who.int/news-room/fact-sheets/detail/human-rights-and-health. Accessed 12 Jan 2021.

World Health Organization. (2024, May 17). *Abortion*. World Health Organization. https://www.who.int/news-room/fact-sheets/detail/abortion

# Chapter 7
# Ethical and Legal Perspectives of Surrogacy in Developed and Sub-Saharan Countries: Case of Australia and Kenya

**Lillian Omutoko**

**Abstract** Surrogacy is a complex and sensitive subject that raises ethical and legal issues. It is a controversial subject whose implications are numerous. Legalization has been proposed as a solution, but it may not be the absolute answer. A possible way out of the intricacies is a well-coordinated system that takes cognizance of the rights of all involved parties. This chapter focuses on the ethical and legal perspectives of surrogacy. Legal and regulatory frameworks governing surrogacy have been discussed with an emphasis on the ethics of surrogacy. The need for surrogacy is occasioned by infertility. Surrogacy is a solution to infertility and an adoption option. It is either altruistic or commercial, depending on the laws of different states. Emerging ethical concerns are the right to dignity, autonomy, justice, and compensation. To deal with emergent ethical issues, some countries have laws that determine the judicial process, and where laws are lacking, couples are left at the mercy of the courts. The nonexistence of laws exposes women and young girls to human rights abuse and exploitation. An internet search was conducted on PUBMED and Google Scholar databases and the research stepwise approach that embraces integration, interdisciplinarity, and transdisciplinarity. The final part of this chapter suggests solutions to the ethical complexities of surrogacy. The solutions include legalization and review of laws, development of guidelines for assisted reproductive technologies, education of interested parties, provision of fertility care, and development of health policies. A framework has been developed to help in the management of surrogacy.

**Keywords** Altruism · Assisted reproductive technology · Infertility · Intended parents · Surrogacy · Surrogate mother

---

L. Omutoko (✉)
Department of Educational Management, Policy and Curriculum Studies, University of Nairobi, Nairobi, Kenya
e-mail: joyce.lillian@uonbi.ac.ke



L. Omutoko, W. Jaoko (eds.), *Bioethics from the Global South*, Advancing Global Bioethics 21, https://doi.org/10.1007/978-3-031-77669-4_7

## 1 Introduction

Infertility is a concern in Africa because of cultural expectations to bear children and social and psychological effects, which makes it a personal, social, and public health issue. As much as infertility is a global issue, the majority of cases are reported in low- and middle-income countries, also known as third world countries (Abebe et al., 2020).

Infertility is a disease of the male and female reproductive systems defined by the failure to achieve pregnancy after 12 months or more of regular, unprotected sexual intercourse. The causes of fertility vary. Male infertility is caused by low sperm production, abnormal sperm production, sexually transmitted infections, blockages that prevent sperm delivery, and health problems such as diabetes, among others. In women, age and lifestyle factors cause infertility (WHO, 2020; Omanwa, 2021). Infertility can be primary or secondary; primary infertility is when a pregnancy has never been achieved, and secondary infertility is when at least one pregnancy has been achieved (WHO, 2020; Sun et al., 2019). Infertility is a health, social, and public health problem; it causes psychological and social effects such as fear, guilt, depression, self-blame, marital stress, emotional abuse, intimate partner violence, divorce, and even loss of social status. Infertility affects individuals, relationships with couples, families, and society. Furthermore, the inability to conceive and get children can affect mental health and well-being (Omanwa, 2021; Chiware et al., 2021).

Infertility rates stand at 48.5 million couples globally, with a prevalence of about 3.5–16.7% in low resource countries and 30–40% in sub-Saharan Africa (Madiziyire et al., 2021; Chiware et al., 2021). It is believed that 10% of global infertility occurs in developing countries. This notwithstanding, the fertility rate has reduced by half over the last 50 years, and the consequences of low fertility are a concern in many countries (Chambers & Fauser, 2021; WHO, 2010). Globally, the prevalence rate of female infertility increased from 1367 per 100,000 in 1990 to 1571 per 100,000 in 2017. Values and trends of burden estimates of infertility in 195 countries indicate that the prevalence of female infertility is relatively higher than that of male infertility (Sun et al., 2019). In Africa, the primary fertility rate is 49.91%; the proportion of infertility in North Africa is 70.56%, West Africa is 41.57%, South Africa is 40%, and East Africa is 30.37%. Secondary fertility rates are 69.63% in East Africa and 29.58% in North Africa (Abebe et al., 2020). The rates of infertility are important for legislation, policy-making, equitable access to fertility care services, and addressing risk factors associated with infertility. This is a human rights issue since individuals and couples have a right to have the number of children they want when they want them (Office of the High Commissioner for Human Rights, 2000).

Due to the high rate of infertility and increasing numbers of people looking for solutions to childlessness, there is a need for the provision of fertility care and the legalization of surrogacy. Couples with infertility issues have an option of adoption of children or seeking solutions from Assisted Reproductive Technologies (ART) to satisfy their right to having children. It is estimated that two in every ten couples face infertility issues, and most of these couples opt for in vitro fertilization (IVF), surrogacy, or adoption. However, an ethical question arises as to whose interest ART satisfies: is it in the parents' interest or the child? Surrogacy is also an option, but an unclear and apparent lack of laws leaves couples at the mercy of the courts. Surrogacy involves an agreement in which a woman consents to bear a child for another person or couple. The parties involved in the agreement are the surrogate mother and the "intended parents" or "commissioning parents." In this chapter, we shall use the term "intending parents." Surrogacy involves the active participation of a group of people, all having human rights, including the child to be born. This is why human rights–based and legal approaches have been used to develop a framework which can be used to reduce the legal and ethical issues that affect individuals, families, and communities (Beer & Bosch, 2020; Australian Human Rights Commission, 2016). Every human being has universal alienable rights. A surrogate arrangement raises concern for the rights of the intended parents, which include the right to privacy, the right of a couple to have children, the right to health, and the right to reproductive health as stipulated in the Bill of Rights in numerous constitutions and declarations (United Nations, 2000, 2014, 2015; Australia's Constitution, 2010; Constitution of Kenya, 2010; The Constitution for the Republic of South Africa, 1996). The surrogate mother has the right to bodily integrity which includes dignity, physical and mental integrity. For a prospective surrogate mother to make a free and informed decision as to whether to be a surrogate mother or not, sufficient regulatory protections need to be in place to protect her from exploitation. Furthermore, power relations have to be taken into account while developing regulatory policies. Surrogacy has the potential to result in exploitation where power imbalances exist. The imbalances are created by unequal relationships between the surrogates and medical doctors and between the surrogates and the intended parents from rich countries. This puts the surrogates in a precarious situation because they have no power to make decisions about the birth process and where they will stay during the pregnancy (Lepapa, 2021). After the pregnancy has started, surrogate mothers are placed in surrogacy homes, thus limiting their freedom and choice. Conditions in these homes are at times deplorable and could compromise human rights. Surrogate mothers have been exploited as they use such arrangements to alleviate poverty. Quite often protections are not provided for them or their children, and they are induced through payments that are not commensurate with the risks they must undergo (https://www.youtube.com/watch?v=rSqR-nE62Dk). In India, surrogates prefer this to working in an industry where they are underpaid and harassed. Thus, surrogacy is an alternative source of income despite the ethical issues and risks involved.

Many African countries lack specific legal frameworks for surrogacy and comprehensive legislation. The absence of clear regulations leads to uncertainty, potential exploitation, and inadequate protection of all involved parties. The knowledge gap regarding surrogacy is multifaceted and presents legal and ethical challenges. A comparative analysis of surrogacy laws in South Africa and Nigeria reveals significant disparities (Adelakun, 2018). Diverse cultures, traditional beliefs, and community norms are likely to transpose with modern legal and ethical environmental considerations which create a gap. There is a need to bridge this gap through legislation, moral discourse, and cultural sensitivity. Due to the lack of clear guidelines and legislation, empirical research is scarce in the African context, which reveals a need for research and collaborative efforts of legal experts, bioethicists, healthcare workers, and policymakers.

Notably, surrogacy laws are influenced by cultural legal and ethical perspectives. Surrogacy was regulated in South Africa in 2007, and it is governed by the Children's Act 38 of 2005. To use a surrogate, parents must comply with the Act's requirements, which include a written agreement with the surrogate mother by the order of the High Court; the surrogate mother and the intending parents must live in South Africa when they enter the agreement, and at least one commissioning parent must be based in South Africa. Psychological evaluations of all parties are also mandatory (https://www.justice.gov.za/legislation/acts/2005; Adelakun, 2018; Burnett, 2024; Preez, 2024). While South Africa has regulations, other African countries do not have legal frameworks, and they may not have surrogacy laws. Surrogacy laws would help in the protection of all parties and ensure fairness by stipulating a process that caters for respect, privacy, confidentiality, and justice. Regarding other African countries, Nigeria has no specific legal framework, but there have been attempts to identify legislative gaps in surrogacy. Reports have revealed the existence of the practice in various forms, such as monetary compensation, cross-border surrogacy, and "baby factories." In this country, there are no legislative measures for governing surrogacy; however, there has been a pending bill since 2012 (Adelakun, 2018). For Kenya, Uganda, and Tanzania, there is limited information on the legislation of surrogacy; therefore there is a need to research each country's legal stance. To provide a better understanding of surrogacy in sub-Saharan Africa and the case studies in question, more insight into Kenya is discussed.

In Kenya, surrogacy operates in a legal and regulatory void. The law is silent on legal parentage in surrogacy arrangements. The absence of a clear regulatory framework and limited literature on the subject, the majority of which is available in the media, compounds the challenge and is a drawback to research (Omanwa, 2021; Lepapa, 2021). The case law in Petition 443 of 2014 brings out the need for a law to regulate surrogate arrangements to protect involved parties, particularly children whose interests should be paramount (http://kenyalaw.org/caselaw/cases/view/105803/). A regulatory framework will define the journey of surrogacy, advise, and protect all parties. It will guide hospitals and fertility clinics in the development of related policies essential for the protection of the rights of surrogates, intended parents, and the child. The cost of surrogacy in Kenya is much lower than it is in other countries, thus making it a preferred destination for international surrogacy (Lepapa, 2021; Rozée et al., 2020; Australian Human Rights Commission, 2016). The Kenya Assisted Reproductive Technology Bill, 2019, proposes the formation of the Assisted Reproductive Technology Authority that will develop standards, regulations, and guidelines on assisted reproduction and establish and maintain a confidential database of persons receiving or providing related services. The bill will regulate the rights and obligations of the parties involved and protect them from exploitation and human rights abuse. Surrogates who have been interviewed have complained of coercion, exploitation, intimidation, human trafficking, forced abortion, identity forgery, and fraud. The bill in Nigeria has been delayed and has not been passed as law. The objectives of this chapter are to (1) describe the legal and regulatory frameworks governing surrogacy, (2) discuss the legal and ethical perspectives of surrogacy, (3) and develop a framework that can be used to reduce the intricacies of surrogacy using human rights and legal- and ethical-based approaches.

## 2 Methodology

The methodology applied is a literary analysis of empirical and quantitative data collected from publications by the World Health Organization (WHO), the United Nations, and journal papers from different countries on surrogate motherhood as well as ethical and legal perspectives of surrogacy. A systematic approach was used in selecting literature on the legalization of surrogacy from 2013 to 2021. The databases used to conduct the Internet search were PubMed and Google Scholar. The global bioethical frameworks and new analysis categories were used to analyze ethical concerns (Ten, 2016; Garzon Diaz, 2021). The new analysis categories allow navigation of the intricate ethical landscape in the contemporary global world. In total, 56 papers were reviewed, and 23 were selected for in-depth review. A 2014 Australian case study that made global headlines on international surrogacy and another one from Kenya were reviewed to show the legal and ethical issues in surrogacy. The papers were analyzed thematically and synthesized to respond to the objectives. The stepwise research process that embraces integration, interdisciplinarity, and transdisciplinarity was applied to understand surrogacy and propose solutions to one of the complex issues affecting human health and life (Hadorn et al., 2008; Segalàs et al., 2014; Zasukhina, 2014).

## 3 Legal and Regulatory Frameworks Governing Surrogacy in Developed and Sub-Saharan Countries

Surrogacy is illegal in some countries, while the law is silent in others. Where there is prohibition, the laws are restrictive to protect the rights, dignity, and welfare of surrogate mothers and children to be. Ethical dilemmas arise when intending parents travel to foreign countries for reproductive markets. Examples of the dilemmas include whether foreign birth certificates should be accepted, whether citizenship should be granted, or whether intending parents should be recognized as legal parents. In such situations, the underlying thought is that the best interest of the child should prevail (Beers & Bosch, 2020). Like human rights, reproductive rights are fundamental freedoms that are an entitlement for women. Reproductive rights give couples and individuals privileges to decide whether to have children, when, and how many. They also have a right to the highest attainable standard of sexual and reproductive health. Various international documents and national laws recognize these rights. Articles 12 and 16 of the Convention on the Elimination of All Forms of Discrimination against Women obligate the states to ensure access to health care services, including those related to family planning. The International Covenant on Economic, Social and Cultural Rights affirms the right to the highest attainable standard of health. Article 12 protects the right to the highest attainable standard of health. The Convention on the Elimination of All Forms of Discrimination against Women and the International Covenant on Economic, Social and Cultural Rights protect women from discrimination of all kinds. Article 24 of the Convention protects the rights of children to the highest attainable standard of health, and Article 14 of the Protocol to the African Charter on Human and People's Rights on the rights of women in Africa refers to the right to sexual and reproductive health (United Nations, 2000, 2014; Office of the High Commissioner for Human Rights, 2000). The International Covenant on Economic, Social and Cultural Rights Article 12.1 recognizes the right of everyone to the enjoyment of the highest attainable standard of physical and mental health. The right to health is dependent on the realization of human rights according to the International Bill of Rights, with human dignity stipulated as one of the rights (Office of the High Commissioner for Human Rights, 2000). The Universal Declaration of Human Rights (UDHR) recognizes that everyone has human rights that must be protected by law, a requirement that is the responsibility of states. While Article 16 of the Charter addresses having a family as a human right, Article 3 of the Universal Declaration of Bioethics and Human Rights (UDBHR) affirms that human rights and fundamental freedoms are to be respected and that the interests and welfare of the individual should have priority over the sole interest of science and society. Article 11 of UDBHR recognizes that human dignity, human rights, and fundamental freedoms should not be violated (United Nations, 2015; UNESCO, 2005). The National Laws of South Africa, the only African country that has legalized surrogacy, enlist the Bill of Rights; the right to human rights, dignity, and healthcare as fundamental rights protected by law. The Constitution of Kenya addresses the right to life under the Bill of Rights, while the Australian Constitution does not have a bill of rights; however, some protections are given by the constitution against legislative or executive action by the Commonwealth. Surrogacy in Australia is regulated at the state and territorial level, which is the second level of government. Australia recently amended the reproductive health laws because of increased legal suits and ethical concerns. The amended health laws do not allow commercial surrogacy and illegalize the undertaking of surrogacy overseas. Despite the amendment, procreative tourism to other countries where laws are less restrictive or absent is still popular, an indication that there could be a need to legalize surrogacy and deal with emanating problems realistically (Piersanti et al., 2021; Lepapa, 2021, Christmas, 2020; Australia's Constitution, 2010; The Constitution of the Republic of South Africa, 1996). Currently, there is no law in Kenya regulating surrogacy arrangements, and the absence of public policy or legislation exposes prospective parents, affected parties, and children to psychological, economic, and social effects. Notably, informal surrogacy has been in existence for a long time, from as far back as the biblical era, and it has been shaped by various traditional cultural and social norms. For example, family members or friends would be motivated by altruism, and they would become surrogate mothers without any formalities. Therefore, surrogacy is a reality that nations should come alive to (http://kenyalaw.org/caselaw/cases/view/105803/). Attempts to legalize surrogacy have been made more complicated by political debates, which leave the practice to be intertwined with medical and physical risks, issues of autonomy, justice, coercion, exploitation, and abuse (Lepapa, 2021; Payne et al., 2020; Mutugi & Lenaola, 2018).

We shall now assess the legislative and regulatory frameworks of various countries for a more comprehensive understanding of legal and ethical issues in surrogacy. Previously, surrogacy was legal and commercialized in India; this encouraged reproductive tourism from other parts of the world by couples who wished for children but were not able to get them (Revel, 2009). Most of those seeking these services were wealthy people from developed countries, a situation that created numerous ethical and legal issues. Literature documents minimal changes that have been made in the law; therefore, a gap exists regarding the current status of surrogacy. The changes made are an attempt to reduce legal and ethical issues and court disputes.

Surrogacy is legal in some countries in Europe and other countries of the world. In some countries, there is no clear regulation by existing laws, and where there are specific laws, there are differences in the judicial approaches. In some states, surrogacy is illegal, while in others, only altruistic surrogacy is legal, as is in India and Australia. In Australia, compensated surrogacy is illegal, and it is a criminal offense to undertake surrogacy overseas (Piersanti et al, 2021; Rozée et al, 2020; Stafford-Bell et al., 2014). Commercial surrogacy is highly controversial, especially in developing countries, when it is performed by local women for wealthy individuals. By 2018, India was considered the top destination for commercial and transnational surrogacy. Scandals in India, Thailand, and Nepal led to the Indian Parliament's decision to convert surrogacy into an altruistic, domestic, and relational practice. Currently, surrogacy can only be carried out by a close Indian relative of a married Indian couple without payment (Rozée et al., 2020; Have, 2016).

Both types of surrogacy are allowable in the United States and Canada; however, different states have different regulations. The same case applies to Australia, where different states have varying regulations. Surrogacy is not illegal in Victoria, South Australia, and Tasmania, but the strict controls and payments relating to it make it impossible for commercial surrogacy to be conducted. Notably, altruistic surrogacy is legal in Australia. Within the 27 member states of the European Union, there is diversity in legislation. In Germany, surrogacy is prohibited because of a violation of morals. The only countries that allow surrogacy in Europe are the United Kingdom, Belgium, the Netherlands, and Finland. In Italy, surrogacy is referred to as “uterus for rental.” Italy allows surrogacy through the Medically Assisted Reproduction Law 40/2004 based on moral justification, which generates ethical debates. Surrogacy has been legal in Greece since 2002, in most of the countries of the former Soviet Union, and in Israel since 1996. The surrogate is paid for legal and insurance expenses as well as compensation for time, loss of income, and pain. On the contrary, African countries have not legalized surrogacy in South Africa, while Asia has had numerous cases of children being born through commercial surrogacy; a case in point is China (Have, 2016). Data on the legalization of surrogacy in sub-Saharan countries is limited; therefore, there is a need for more research on surrogacy. Despite the absence of legalization, there is a need to know the extent to which fertility care and surrogacy are practiced to deal with legal and ethical issues since they exist (Mutugi & Lenaola, 2018). All Muslim countries prohibit surrogacy by law and professional medical codes of ethics (Piersanti et al., 2021; Have, 2016).

The increased demand for fertility services such as in vitro fertilization (IVF), surrogacy, and adoption, as well as emanating legal and ethical concerns worldwide, show the need for enabling laws and policies to regulate third-party reproduction to ensure universal access without discrimination and the protection and promotion of human rights (WHO, 2020; Omanwa, 2021). To the best of my knowledge, the literature reviewed from 2018 to 2021 covers the legalization of surrogacy but does not cover the reviewed laws except for India, which reviewed relevant laws in 2018, and Australia. In Australia, compensated surrogacy is illegal, and it is a criminal offense to undertake surrogacy overseas; therefore, there is a need to review the laws to establish changes that have been made to address the emanating ethical issues that have bridled the practice (Stafford-Bell et al., 2014; Rozée et al., 2020).

Despite the proposal by several authors that regulation is the answer to the ethical and legal issues emanating from surrogacy, inquiries during the law reforms in Australia reveal that regulation of surrogacy in Australia and possibly in other countries does not quite prevent surrogacy but rather leads to more secrecy with negative consequences for children. This calls for the introduction of a proper system of surrogacy, and this is where the proposed framework for management of surrogacy comes in (South African Law Reform Institute (SALRI, 2018). SALRI reviewed the Family Relations Surrogacy Amendment Act 1975, which regulates surrogacy in South Australia, occasioned by the need for a standalone surrogacy act. It proposes that the framework will be multi-thronged, it will use the human rights–based approach, define the surrogacy journey, and propose the stakeholders who should be involved in ensuring protection of all parties. We shall now use the cases of Australia and Kenya to discuss the legal and ethical perspectives of commercial surrogacy. This section will show the existence and commonality of legal and ethical issues in countries where surrogacy is legal and where there is no legislation. Two case studies have been used to compare the legal and ethical issues in developed and sub-Saharan countries. Case studies were drawn from Australia and Kenya; the Gammy case was identified because it is a perfect case of international surrogacy, while the case of Monica illustrates how commercial surrogacy affects women in developing countries.

## 4 Legal and Ethical Perspectives of Commercial Surrogacy: The Case of Australia and Kenya

Surrogacy is delicate and controversial, it is a situation where the surrogate mother agrees to have a baby for another person or a couple who are not able to get a baby. Once the baby is born, the surrogate gives up custody of the baby to the intended parents. Some authors refer to such a baby as a "team baby," which raises parenthood questions and exposes intended parents, born children, and egg donors to unsafe clinical practices and undesirable ethical and legal practices (SALRI, 2018). National laws, the UDHR, UDBHR, and CEDAW which stands for the Convention on the Elimination of All forms of Discrimination Against Women recognize human rights as belonging to everyone. These rights include the right to dignity, the right to health, reproductive rights, the right to autonomy, justice, and compensation, the right to enjoy the highest attainable standard of physical and mental health, and the right to have children they want and when they want them (United Nations, 2000, 2014, 2015). In the case of Australia, there is an argument that there should be no distinction between altruistic and commercial surrogacy because both can be exploitative with the possibility of coercion regarding the extent a woman exercises her free will to become a surrogate. This would be through undue pressure from family members or friends on behalf of those desperate to have a child. Surrogacy may be exploitative in developing countries where surrogates may be poor, illiterate, and uninformed about their rights (SALRI, 2018). The practice in Australia is a good contrast to Kenya's, with the practice being legal in Australia and illegal in the latter. Risks that women and young girls face have been exposed. There are risks of exploitation due to the socioeconomic status of a majority of women in developing countries. Such women are attracted by financial incentives, which are sometimes noncompensatory. In Kenya, compensation is not commensurate with the risks you are likely to encounter. The perceived benefits mar autonomous choices and instead draw the women into a coercive relationship with the intended parents and expose the former to potential exploitation and abuse (Beers & Bosch, 2020; Piersanti et al., 2021; Lepapa, 2021).

Several court cases have arisen in various countries of the world. I will use the Australian Gammy case to illustrate possible legal and ethical issues in commercial surrogacy. The ethical concerns in commercial surrogacy would be privacy, confidentiality, informed consent of the donor(s), and beneficence concerning the recipient. Beneficence brings up the issues of compensation versus risks of the third party, that is, the surrogate. A question that remains unanswered is what is the monetary value of carrying a baby for someone else and then having to detach physically and emotionally from a 9-month bond? There are risks of the surrogate mother refusing to give up the child or the commissioning parents refusing to accept the child (Mutugi & Lenaola, 2018; Lepapa, 2021).

The case of the Australian couple raised hue and cry. David and Wendy commissioned a Thai woman through a surrogacy agency and delivered twins. One of the twins had Down syndrome, and the Australian couple abandoned him and took his twin sister to Australia. The couple claimed that they did not know about the existence of the boy, but the staff at the agency said that Mr. and Mrs. Farnell knew about the existence of both children and had stayed with them at the hospital in the same room for 4 weeks after birth. The agency employee said that they did not know what to do with Gammy when they realized that he had Down syndrome. Doctors suggested an abortion but changed their minds on realizing that the life of the other twin would be endangered. The surrogate mother was clear that she wanted to have the baby, and she carried the pregnancy to term. The couple maintained that they did not leave Gammy behind because he had Down syndrome but that they had been told that he had a heart condition and he would not survive after his birth (https://www.couriermail.com.au/lifestyle/david-farnell-father-of-baby-gammy-and-convicted-sex-offender-dies/news-story). This case sparked an international surrogacy scandal over surrogacy laws, and it prompted investigations into the commercial surrogacy industry in Thailand. It was revealed that Farnell was a convicted sex offender; he had been jailed for 3 years for sexual offenses against girls in Bunbury. He had been fighting an asbestos-related illness. When Chanbua, the surrogate mother, found out that Farnell was a child sex offender, she applied to the court to have Gammy's sister, Pipah, returned to her in Thailand, but the judge determined that Pipah should not be removed from the only family she has ever known and that Farnell was at low risk of offending. Justice Stephen Thackeray found that Farnell and his wife had not abandoned Gammy but that Chanbua had "fallen in love" with the twins she was carrying and had decided to keep the boy. Gammy, now named Grammy, still lives in Thailand and began attending kindergarten in 2017 and is reportedly doing well (Mutugi & Lenaola, 2018; Christmas, 2020; https://www.couriermail.com.au/lifestyle/david-farnell-father-of-baby-gammy-and-convicted-sex-offender-dies/news-story). The principles and values being transgressed here are human dignity and human rights of the twins. The ethical principles violated are respect for persons, beneficence, and justice. The twins had no place in the decision-making process, the commissioning parents made decisions that did not consider the interests of the children. The laws in Australia only permit altruistic surrogacy, this is why the couple traveled to Thailand where the laws were a bit lenient in 2014, before the Gammy case. Until then, "'reproductive tourism'" and surrogacy agreements were deemed void as against public policy and therefore they were not regulated by law. Despite well-written contracts, disputes relied on judges' decisions as to who between the commissioning parents and the surrogate mother was the legal parent of the child; no wonder Gammy remained in Thailand and Pipah was taken to Australia (Stasi, 2017). Autonomy is violated, although it can be argued that the twins were minors; however, taking one of them and leaving another in Thailand deprives them of sibling love and the right of growing up together. The question of beneficence also arises. To whose benefit was this process intended? Was it for the children, the surrogate mother, or the commissioning parents? Viewed from the perspective of benefit versus risk, the children had no benefit because they were separated at birth and Pipah was being brought up by a father who had been jailed because of involvement in crime. The surrogate mother could have benefited from compensation, but the case revealed that she was not fully compensated. Again, the question is how much compensation would be commensurate with the burden of pregnancy and physical and emotional pain? This analysis shows that Mr. and Mrs. Farnell were to benefit more by having a third party carry the pregnancy for them, then they take over the parenthood. With regards to justice, taking Pipah to Australia and leaving Gammy in Thai was not fair, even though he had Down syndrome. When Chanbua, the surrogate, realized that Mr. Farnell was a convict, she requested to be given Pipah on grounds of risk, it was also unfair to separate the twins and deny them warm brotherhood. Ultimately, the twins were not protected from harm by this arrangement between the commissioning parents, the surrogate mother, and the state, who would have provided protection through legislation that prevents complex situations such as this (Christmas, 2020; Zasukhina, 2014). Full disclosure was also flouted at the surrogacy agency; the law should demand for couples to disclose their history and health. Gammy may have had Down syndrome because Mr. Farnell had been suffering from an asbestos-related disease, which could have been fatal. Therefore, there was a need for disclosure so that the children do not have to be at risk of growing up without parents, although death is unpredictable. After a series of high-profile international scandals drew public attention to women exploitation, human trafficking, and legal ambiguities surrounding surrogacy in Thailand, in 2014, Thai authorities launched a crackdown on commercial surrogacy, and new legislation was made to regulate commercial surrogacy (Stasi, 2017).

It is acknowledged that different cultures hold diverse moral values and ethical norms. Some communities in Kenya practice traditional surrogacy. The Kamba community of Eastern Kenya has provision for surrogacy. The surrogate has the arrangement to carry and deliver a child for another person who is unable to have a child. In such arrangements, she is the egg donor and biological mother of the child. Such situations call for understanding and respecting cultural diversity without compromising ethical principles; therefore, the need for culturally responsive approaches to surrogacy (Ten, 2016). Surrogacy also exists in the form of wombs for hire. The case of Monica—not her real name—is not publicized due to the secrecy that embroils surrogacy in countries where there is either no legislation or the law is silent. The case was published in *Elephant*—a media magazine (https://www.theelephant.info/investigations/2021/05/28/hard-labour-the-surrogacy-industry-in-kenya-part-i/). In developing countries, surrogacy is commercial, and women with low socioeconomic status see it as a source of income that can save them from poverty. Women who carry pregnancies on behalf of others and the born children are vulnerable in various ways. The "payment" they get is more of an inducement than compensation. With the absence of clear laws, commercial surrogacy agencies take responsibility for arrangements with intended parents and surrogates.

Monica was a single mother of two earning $55 per month as house help before she lost her job. While looking for a job, she found what she thought was a life-changing opportunity. She joined a WhatsApp group seeking surrogates and then met a surrogate recruiter. The recruiter provided Monica with bus fare to meet him in Nairobi City, and he offered to pay her $5,700 for the entire process. She was to be paid $180 per month and $4100 after delivery, which is good income for someone who previously earned $55 per month. After the meeting, Monica was to meet the recruiter at an IVF clinic for tests and hormonal medication to prepare her uterus for pregnancy. All the tests and medication were done and given without obtaining her consent. Verbal consent was obtained later as an afterthought, and no contract was signed. The embryo was implanted in her womb, and she was taken to a hostel. The interests of Monica and the child to be born were never taken into account. What mattered most was a baby for the intended parents, who were foreigners, most likely from a rich country. While at the hostel, the recruiter controlled her life, including what she ate and when and where she traveled. Her rights were infringed, and she was denied the opportunity to make an informed decision as to whether to participate or not (https://www.theelephant.info/investigations/2021/05/28/hard-labour-the-surrogacy-industry-in-kenya-part-i/).

Six months into the pregnancy she was to travel out of the country where the intended parents of the baby lived. Monica was taken to the immigration office to apply for a passport, and within a few days, she was on her way out of Kenya. She was booked in a hotel where she lived for 2 months, and the intended parent would visit her. In the eighth month of pregnancy, the intended parent was to pick her up and take her to live with his parents, where she would get better care. But when he arrived, he explained that he had changed his mind about having the baby because his parents didn't want the baby. Monica was checked into another hotel, and a doctor injected her with a drug that caused severe contractions. She was then taken to a back-street abortion, and she went into labor for 12 h. When the baby failed to come out, she was transferred to another hospital. A cesarean section was conducted, but the baby had already died. This was unjust treatment considering what she had gone through and the sacrifices made to earn some money. It is a case of human rights abuse and exploitation. She was not allowed to make any autonomous decisions. Instead, decisions were made for her by "the agency," the intended parent, and the doctor. This encounter emphasizes the risk of exploitation and abuse of surrogates due to the lack of regulation. (https://www.theelephant.info/investigations/2021/05/28/hard-labour-the-surrogacy-industry-in-kenya-part-i/).

Surrogacy is a reality that can no longer be ignored. The emanating challenges can only be resolved through regulation and monitoring of the practice to protect vulnerable parties from exploitation and harm. Removing surrogacy from obscurity would clear the way for a transparent procedure from which families can be comfortably built. Legalization and regulation would also discourage families from traveling abroad to avoid the law in their own countries (Beers & Bosch, 2020). The proposed framework will help in introducing transparency and systematizing the process.

## 5 Framework to Reduce the Complexities of Surrogacy Using the Human-Based Approach and Best Practices for Surrogacy

The two case studies illustrate some of the risks that surrogates and children to be born are exposed to due to the lack of clear laws and restrictions that drive interested parties to potentially exploitative arrangements. A legal and ethical approach to surrogacy could be the solution to this global problem. Setting up laws, regulations, and respective national authorities or overarching bodies to oversee the process would be helpful. The authority can be engaged in the process from an early stage before conception, which means that the pregnancy will not commence unless a contractual agreement has been signed between intending parents and the prospective surrogate mother. If it is the court giving permission, then complying with this procedure gives the parents legal parenthood. The ethical and legal approach can be used alongside the proposed framework.

A framework is proposed for developing a procedure for managing surrogacy and to help in alleviating the suffering. A human rights–based approach has been used to show what states, the legal fraternity, health personnel, and stakeholders should work together to protect the interests of all parties. Building the framework requires information on who should be involved and what should be considered in surrogacy arrangements.

The human rights approach applies in the sense that the rights and welfare of the born child, the rights, well-being, and physical wellness of surrogates, and the interests of intending parents should be catered for. The definition of rights should be based on the UDHR, UDBHR, CEDAW, and national laws, which address human rights and human dignity, the Bill of Rights, and reproductive rights, specifically the right to health. National laws enhance the declarations and therefore can be used to protect parties (United Nations, 2000, 2014, 2015; UNESCO, 2005; Australia's Constitution, 2010; Constitution of Kenya, 2010). The framework will guide the surrogacy process and prevent situations whereby individuals and fertility clinics are not aware of the requirements for eligibility and residency (Law Commission and Scottish Law Commission, 2019; SALRI, 2018). A clear process will demystify surrogacy and help in reducing social stigma (Lepapa, 2021; Beers and Bosch, 2020). Figure 7.1 illustrates how the complications of surrogacy can be systematically handled by states, authorities, stakeholders, and organizations.

This chapter proposes the use of a bottom-up approach to align priorities with stakeholder needs and create a holistic strategy. The approach necessitates definition of the journey of surrogacy and availing information on what should be considered

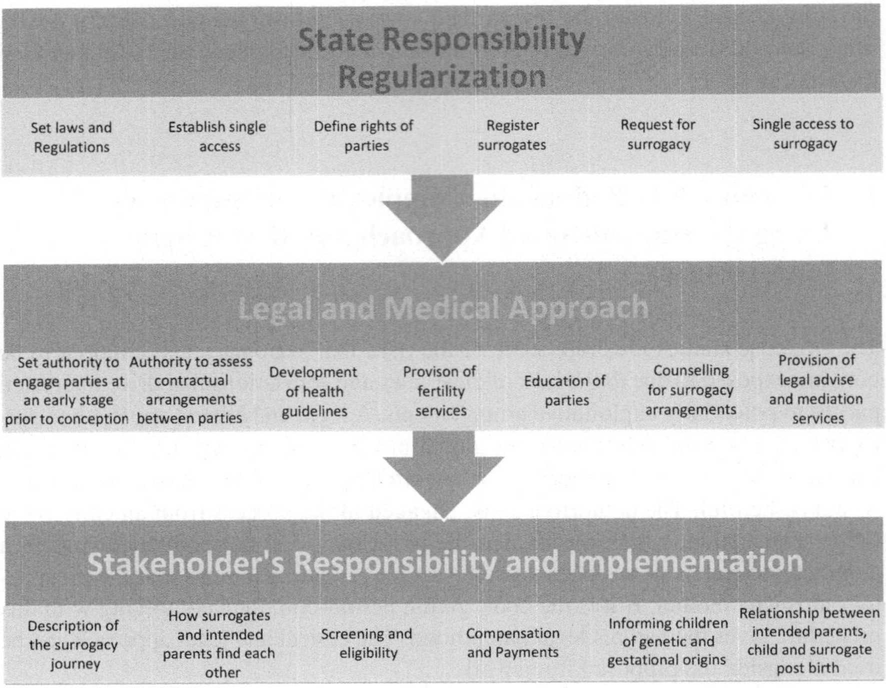


**Fig. 7.1** Regulation of legal and ethical surrogacy for the best interest of all involved parties

in surrogacy arrangements such as the establishment of single-access agencies, definition of rights of parties, and a surrogate's register. This would make the ensuing law easier to implement, improve access to lawful surrogacy and protect intending parents and the children. Country-specific frameworks can be developed based on the respective Bills of Rights, Reproductive Health Bills, laws, and regulations where they exist (Law Commission and Scottish Law Commission, 2019; SALRI, 2018).

## 6 Conclusion

Infertility is a natural problem that makes couples and those with certain health problems seek surrogacy as a solution. Surrogacy enables childless couples and single parents to have the most desired children; however, it continues to generate debates the world over. It is concluded that surrogacy can be solved by human-based and legal and ethical approaches that encompass the development of national laws and regulations at state and institutional levels, the provision of fertility care, and the development of relevant health policies and guidelines for assisted reproductive technologies. Legalization and review of existing laws will help in regulating the industry, while education and sensitization will equip interested parties in making informed decisions. Homegrown remedies such as herbal medicine, traditional surrogacy, and polygamy in communities that accept the practices have a place in meeting the need for having children for families who would otherwise not have children naturally.

## References


Abebe, M. S., Afework, M., & Abebe, M. S. (2020). Primary and secondary infertility in Africa: systematic review with meta-analysis. *Fertility Research and Practice, 6*, 20. https://doi.org/10.1186/s40738-020-00090-3

Adelakun, O. S. (2018). The concept of surrogacy in Nigeria: Issues, prospects and challenges. *African Human Rights Law Journal, 18*, 605–624. https://www.saflii.org/za/journals/AHRLJ/2018/28.pdf

Australian Constitution. (2010). Australian Constitution. Retrieved from https://www.legislation.gov.au/Details/C2010C00001

Australian Human Rights Commission. (2016). *Legislative aspects of surrogacy arrangements*. Australian Human Rights Commission to the House of Representatives Standing Committee on Social Policy and Legal Affairs. https://humanrights.gov.au/sites/default/files/20160217_AHRC_SurrogacySubmission.pdf

Beers, B. V., & Bosch, L. (2020). A revolution by stealth: A legal ethical analysis of the rise of pre-conception authorization surrogacy arrangements. *The New Bioethics, 26*(4), 351–371. https://www.tandfonline.com/doi/pdf/10.1080/20502877.2020.1836464?needAccess=true

Burnett Attorneys and Notaries Inc. (2024). *Surrogacy laws in South Africa*. https://www.burnettlaw.co.za/surrogacy-laws-in-south-africa/

Chambers, G. M., & Fauser, B. C. J. M. (2021). Access to ART treatment and gender equality. *Reproductive Biomedicine Online, 42*(4), 687–690. https://www.rbmojournal.com/article/S1472-6483(21)00097-3/full text

Chiware, T., Vermeulen, N., Farquharson, R., Kiarie, J., Lundin, K., Matsaseng, T. C., Ombelet, W., & Toskin, I. (2021). IVF and other ART in low-and middle-income countries: A systematic landscape analysis. *Human Reproduction Update, 27*(2), 213–228. https://doi.org/10.1093/humupd/dmaa047. https://watermark.silverchair.com/dmaa047.pdF.

Christmas, P. (2020). *Father of Baby Gammy and convicted WA sex offender David Farnell dies*. https://7news.com.au/news/wa/father-of-baby-gammy-and-convicted-wa-sex-offender-david-farnell-dies-c-1168612

Constitution of Kenya. (2010). Kenya Gazette, CXII (88), 3 September 2010. Retrieved from https://new.kenyalaw.org/akn/ke/act/2010/constitution

Garzon Diaz, F. A. (2021). New analsys categories of global bioethics. *Revista Latinoamericana de Bioética, 21*(2), 7–10. https://doi.org/10.18359/rlbi.5913

Have, T. H. (2016). *Surrogacy in book: Encyclopedia of global ethics*. https://doi.org/10.1007/978-3-319-05544-2_409https://www.researchgate.net/publication/299947629_Surrogacy

Hadorn, H., Hoffmann, G., Riem, H., Klemm, B., Grossenbacher, S., Mansuy, W., Joye, D., Pohl, C., & Zemp, E. (Eds.). (2008). *Handbook of transdisciplinary research*. Springer.

Law Commission and Scottish Law Commission. (2019). *Building families through surrogacy: A new law*, pp.49–62. https://s3-eu-west-2.amazonaws.com/lawcom-prod-storage 11jsxou24uy7q/uploads/2019/06/Surrogacy-consultation-paper.pdf

Lepapa, N. (2021). Hard labour: The surrogacy industry in Kenya- Part 1. *The Elephant*. African Analysis, opinion and investigation. https://www.theelephant.info/investigations/2021/05/28/hard-labour-the-surrogacy-industry-in-kenya-part-i/

Madiziyire, M. G., Magwali, T. L., Chikwasha, V., & Mhlanga, T. (2021). The causes of infertility in women presenting to gynaecology clinics in Harare, Zimbabwe; A cross sectional study. *Fertility Research and Practice, 7*(1). https://doi.org/10.1186/s40738-020-00093-0

Mutugi, M., & Lenaola, I. (2018). *Bioethics of medical advances and genetic manipulation: Legal, philosophical and moral perspectives.* Longhorn.

Office of the High Commissioner for Human Rights. (2000). CESCR General Comment No. 14: The Right to the Highest Standard of Health (Art.12). https://www.refworld.org/pdfid/4538838d0.pdf

Omanwa, K. (2021). *Couples fight for embryos in divorce cases.* https://www.businessdailyafrica.com/bd/lifestyle/society/couples-fight-court-for-embryos-3363694

Payne, J. G., Korolczuck, E., & Mezinka, S. (2020). Surrogacy relationships: A critical interpretative review. *UPSALA Journal of Medical Sciences, 125*(2), 182–191. Taylor and Francios Group. https://www.tandfonline.com/doi/pdf/10.1080/03009734.2020.1725935?needAccess=true. https://doi.org/10.1080/03009734.2020.1725935

Piersanti, V., Consalvo, F., Signore, F., Del Rio, A., & Zaami, S. (2021). Surrogacy and "procreative tourism". What does the future hold from the ethical and legal perspectives? *Medicina, 57*, 47. https://www.ncbi.nlm.nih.gov/pmc/articles/PMC7827900/

Preez, M. (2024). *Surrogacy and the legal process.* https://www.mblh.co.za/OurInsights/ArticleDetail.aspx?Title=Surrogacy-and-the-Legal-Process

Revel, A. (2009). Current status of assisted reproductive techniques (ART). In F. Simonstein (Ed.), *Reprogen-ethics and the future of gender* (International Library of Ethics, Law, and the New Medicine) (Vol. 43). Springer. https://doi.org/10.1007/978-90-481-2475-6_2

Rozée, V., Unisa, S., & de La Rochebrochard, E. (2020). The social paradoxes of commercial surrogacy in developing countries: India before the new law of 2018. *BMC Women's Health, 20*, 234. https://doi.org/10.1186/s12905-020-01087-2

Segalàs, J., Tejedor, G., & Cebrián, G. (2014). (Re)Shaping knowledge: The contribution of Sustainability Science. In GDEE (Ed.), *The global engineer in Sustainable Human Development.* Retrieved June 19, 2019, from https://www.academia.edu/19739065/The_global_engineer_in_Sustainable_Human_Development._Re_Shaping_knowledge_the_contribution_of_Sustainability_Science

South Australian Law Reforms Institute. (2018). *Surrogacy: A legislative framework. A review of Part 2B of the family relationships Act 1975(SA).* https://law.adelaide.edu.au/system/files/media/documents/2019-02/salri_surrogacy_report_oct_2018_0.pdf

Stafford-Bell, M. A., Everingham, S., & Hammarberg, K. (2014). Outcomes of surrogacy undertaken by Australians overseas. *The Medical Journal of Australia, 201*(6), 330–333. https://doi.org/10.5694/mja14.01086. https://www.researchgate.net/publication/265649050_Outcomes_of_surrogacy_undertaken_by_Australians_overseas

Stasi, A. (2017). Protection for children born through assisted reproductive technologies Act, B.E.2558: The changing profile of surrogacy in Thailand. *Clinical Medicine Insights: Reproductive Health., 11*, 1–7. https://doi.org/10.1177/1179558117749603. https://journals.sagepub.com/doi/10.1177/1179558117749603

Sun, H., Gong, T., Jiang, Y., Zhang, S., Zhao, & Wu, Q. (2019). Global, regional and national prevalence and disability-adjusted lie-years for infertility in 195 countries and territories, 1990-2017: Results from a global burden of disease study, 2017. *Aging, 11*(23), 10952–10991. https://www.ncbi.nlm.nih.gov/pmc/articles/PMC6932903/pdf/aging-11-102497.pdf

Ten Have, H. (2016). *Global bioethics: An introduction* (1st ed.). Routledge. ISBN: 9781315648378. https://doi.org/10.4324/9781315648378

The Constitution of the Republic of South Africa, 1996. https://dearsouthafrica.co.za/wp-content/uploads/2018/05/SAConstitution-web-eng.pdf

UNESCO. (2005). *Universal declaration on bioethics and human rights.* http://portal.unesco.org/en/ev.php-URL_ID=31058&URL_DO=DO_TOPIC&URL_SECTION=201.html

United Nations. (2000). *Convention on the elimination of all forms of discrimination against women.* https://www.un.org/womenwatch/daw/cedaw/cedaw27/tun3-4.pdf

United Nations. (2014). *Reproductive rights are human rights. A handbook for national human rights institutions*. https://www.ohchr.org/documents/publications/nhrihandbook.pdf
United Nations. (2015). *Universal declaration of human rights*. https://www.un.org/en/udhrbook/pdf/udhr_booklet_en_web.pdf
WHO. (2010). Mother or nothing: The agony of infertility. *WHO Bulletin*, *88*(12), 881–882. http://www.who.int/reproductivehealth/publications/infertility/en/index.html. Accessed 23 Mar 2014.
WHO. (2020). *Infertility*. https://www.who.int/news-room/fact-sheets/detail/infertility
Zasukhina, V. (2014, July). *Bioethics in transdisciplinary and interdisciplinary context of post-nonclassical science*. Retrieved June 19, 2019, from https://doi.org/10.12955/cbup.v2.464

# Chapter 8
# Ethics of Surrogacy: A Review of the African Perspective

**Emmanuel E. Okenwa-Vincent**

**Abstract** Surrogacy, typically a legally binding contractual agreement where a surrogate carries and delivers a child for a commissioning party, has emerged as a notable fertility solution in Africa. This practice has financially benefited many surrogate families but has also ignited numerous ethical and societal dilemmas among African scholars due to the lack of comprehensive guidelines and regulations. This chapter examines the ethical implications of surrogacy as presented in published journal articles. A detailed search of six databases using specific search terms yielded ten relevant journal articles. These articles were analyzed and thematically categorized to reflect the ethical and moral principles being challenged. The findings reveal significant ethical challenges, including physical, emotional, and psychosocial consequences, often overshadowed by material and socioeconomic motivations. The chapter concludes that these challenges are largely due to inadequate legal and regulatory frameworks and a decline in societal, religious, moral, and cultural values surrounding fertility. In response, this chapter proposes a tripartite approach—comprising legislation and regulation, cultural and religious integrity, and the promotion of ethical literacy—to address the complex issues associated with surrogacy in Africa.

**Keywords** Surrogacy · Surrogacy in Africa · Ethics of surrogacy · Fertility regulations · Ethical principles · Personalism

## 1 Introduction

Surrogacy, a multifaceted bioethical issue, involves a legally binding agreement where a woman, known as the surrogate, carries and delivers a child for another individual or couple, referred to as the commissioning party (Ten Have, 2016). This

E. E. Okenwa-Vincent (✉)
Department of Optometry and Vision Sciences, Kaimosi Friends University, Kaimosi, Kenya
e-mail: eokenwa@kafu.ac.ke

 
L. Omutoko, W. Jaoko (eds.), *Bioethics from the Global South*, Advancing Global Bioethics 21, https://doi.org/10.1007/978-3-031-77669-4_8

arrangement, rooted in deep socioethical complexities, varies significantly across regions in both intent and implementation. Surrogacy agreements manifest in two primary forms: altruistic and commercial. Altruistic surrogacy, observed in countries like South Africa, typically involves surrogates motivated by goodwill, not receiving payment beyond the reimbursement for medical expenses and related inconveniences. This form is often driven by the surrogate's empathetic response to the commissioning couple's medical inability to conceive, as diagnosed by a fertility specialist. Conversely, commercial surrogacy, which is more prevalent and notably practiced in countries like Kenya, Nigeria, Ghana, and so forth, involves financial compensation for the surrogate that substantially exceeds basic medical expenses. Surrogates in such arrangements can earn between 10 and 15 thousand US dollars (Adelakun, 2018). Furthermore, surrogacy can be classified based on the biological connection to the child: partial (genetic) surrogacy involves the male partner of the commissioning couple fertilizing the surrogate's egg through natural or artificial means, while full (gestational) surrogacy uses an embryo created from the commissioning couple's genetic material, implanted into the surrogate's womb through in vitro fertilization (Adelakun, 2018).

The legality of surrogacy varies globally, with some nations having robust regulatory frameworks, while others, particularly many African and other developing countries, exhibit significant legal gaps. For instance, South Africa's Children's Act of 2005 requires that all surrogacy agreements be validated by the High Court before fertilization and mandates a genetic link between the child and at least one commissioning parent (Dawes, 2009). Conversely, in countries like Nigeria, there is an absence of specific legal statutes addressing surrogacy, leading to potential risks such as inadequate medical screening, substandard care, and insufficient psychosocial support for surrogates. Nations such as the United States, India, Thailand, Ukraine, and Russia are well known for their surrogacy arrangements. However, the global landscape is shifting, with countries like Mexico, Nepal, Poland, and Georgia emerging as new centers, despite the lack of international consensus or guidelines (ICCA, 2019). Within Africa, the increasing restrictions in Asian countries have made nations like Nigeria, Kenya, and Ghana attractive destinations for surrogacy. Yet, the lack of stringent regulations in these countries poses significant ethical and legal challenges.

The ethical concerns surrounding surrogacy are profound, particularly in Africa, where societal pressures around infertility amplify the complexities. The commercialization of surrogacy raises critical bioethical questions concerning the commodification of women's bodies, the exploitation of vulnerable women, and the moral and ethical handling of early human life (Ten Have, 2016). These issues, coupled with the evolving legal landscapes, render surrogacy a pressing bioethical issue that demands rigorous scholarly examination and robust legislative action. This chapter aims to critically review the existing literature on the perceptions and ethical considerations of surrogacy in Africa, highlighting the unique challenges and proposing pathways to address these concerns effectively.

## 2 Methodology

The review conducted in this chapter is based on a critical appraisal of journal articles that focus on surrogacy. A systematic search was conducted across six databases: PubMed, Google Scholar, Scopus, JSTOR, Mendeley, and the Web of Science. Using the key phrases "ethics of surrogacy in Africa" and "African ethical perspectives on surrogacy," we retrieved a total of 41 articles. Of these, only ten journal articles—published in the last 10 years and available in PDF format—met the inclusion criteria and were selected for further review. We excluded articles that were abstracts only, not related to the ethics of surrogacy or the African perspective on surrogacy, or not written in English. After this selection process, a comprehensive database of the selected articles was created. These articles were then categorized based on their relevance and importance to the overarching topic of this review.

## 3 Discussions

As outlined in the introductory section of this chapter, surrogacy is a legally binding arrangement in which a surrogate carries a pregnancy for the intended parents to address fertility concerns. Particularly in commercial surrogacy, the surrogate often receives significant financial compensation, which may help alleviate poverty. However, this arrangement is laden with ethical complexities. In many developing countries, including several in Africa, surrogacy has become a fundamental element of what is known as "fertility tourism." Individuals or couples from regions where surrogacy is either banned or economically prohibitive frequently travel to developing countries with more permissive regulatory frameworks concerning fertility treatments. These individuals aim to have children at costs significantly lower than what they would incur in their home countries. From an ethical standpoint, surrogacy presents several critical questions: Is the practice of surrogacy inherently morally reprehensible? If not, does it provide avenues for unethical or exploitative practices? Moreover, which specific aspects of surrogacy are considered ethically questionable? This review seeks to delve deeper into these moral inquiries, focusing on the ethical dimensions of surrogacy. It starts with a thorough examination of the challenges presented by surrogacy, particularly through the perspective of African experiences, and emphasizes commercial surrogacy to provide clarity on the subject.

### *3.1 Problem of Surrogacy in Africa*

Surrogacy, a rapidly evolving facet of assisted reproductive technologies (ARTs), presents complex challenges across Africa, shaped by diverse sociocultural, religious, and regulatory factors influencing its perception and implementation (Alabi,

2021). In countries like Nigeria, Ghana, Kenya, and South Africa, the experiences and challenges of commercial and transborder gestational surrogacy vary significantly, marked by their unique problems and potential solutions (ISF, n.d.). The prominent challenge within these countries remains the absence of clear regulations and policies guiding surrogacy practices. However, South Africa has made significant progress in legislating and regulating surrogacy, empowering courts to approve all surrogacy contracts, which helps protect the rights of all stakeholders and promotes surrogacy as a viable treatment for infertility (ICCA, 2019; Pheme et al., 2020, p. 202). In contrast, countries such as Nigeria, Kenya, and Ghana face substantial legislative and regulatory gaps (Adelakun, 2018; Alabi, 2021; Rashid, 2018). These deficiencies contribute to unethical practices, such as the emergence of baby factories in Nigeria, which result in increased abuse and exploitation of surrogates (Adamu, 2020; Adelakun, 2018). While Kenya has constitutional provisions for surrogacy, it lacks effective implementation frameworks, leading to the proliferation of illegal surrogacies and challenges in realizing constitutional rights (Rashid, 2018). Ghana, meanwhile, presents a mix of challenges and opportunities; despite scarce regulations for ARTs, it has become a preferred destination for transborder fertility treatments due to easy access to surrogates and donors, cost-effectiveness, and fewer regulatory hurdles (Gerrits, 2018). However, this trend may not accurately reflect the local perceptions and experiences of surrogacy in Ghana, as witnessed elsewhere in Africa (Adamu, 2020; Adelakun, 2018; Alabi, 2021; Makinde et al., 2016; Ogar et al., 2018), underscoring the need for more research into the cultural, societal, and ethical contexts of surrogacy in Ghana.

Religious and sociocultural influences constitute another significant problem, often shaping the decision-making process in fertility-related matters (Afolabi, 2018). This influence is starkly observed in Nigeria's Yoruba ethnic group and South Africa's Limpopo province (Alabi, 2021; Pheme et al., 2020). Among the Yoruba ethnic group in Nigeria, Alabi (2021) observed that religious and sociocultural structures influence surrogacy, a highly stigmatized and complex process, and direct their reproductive-related decision-making process. This is also corroborated in South Africa, where Pheme et al. (2020) found that childless South African women of Limpopo province, who would readily uptake surrogacy, reported that culture and religion largely influenced their decisions to uptake surrogacy. With regard to religion, Ogar et al. (2018) wrote that most cultures within sub-Saharan Africa resent the idea of surrogacy because it is biblically wrong. Citing the biblical story of Abraham and Sarah, Ogar et al. (2018) argued that God frowned at the fact that Abraham's wife, Sarah, still infertile in her old age, gave her maid Hagar to the husband to conceive and bear him a child. He further argued that the consequence of this act, viewed as a demonstration of impatience with the Lord, is the cause of Israel and Arab conflicts, even to the present day (Ogar et al., 2018). The multifaceted challenges of surrogacy in Africa call for comprehensive discussions on morality, ethics, and legality, involving a wide array of stakeholders (Umeora et al., 2014). Addressing these issues effectively requires enhancing regulatory frameworks, providing psychosocial support to infertile women, and eradicating exploitative practices. Although surrogacy poses formidable challenges, the South African model demonstrates how legislative and judicial efforts can safeguard the rights and welfare of all parties involved (ICCA, 2019). Given the significant role of surrogacy in aiding childless couples, it is crucial that these concerns are addressed not only from a medical standpoint but also from social and public health perspectives (Pheme et al., 2020).

Furthermore, promoting awareness of surrogacy in communities, especially among infertile women in Africa, where surrogacy is largely misunderstood and stigmatized, is critical (Afolabi, 2018; Alabi, 2021). Insufficient knowledge about surrogacy fosters misinformation, complicating decision-making for potential surrogates and intended parents and heightening the risk of exploitation, particularly in regions with lax regulatory oversight like Nigeria, Kenya, and Ghana (Adelakun, 2018; Rashid, 2018). As the global fertility market increasingly exploits these regulatory gaps, fostering cross-border surrogacy, it is imperative to address the associated risks, particularly for surrogates and children born from these arrangements, who are potentially put at risk of identity loss (Makinde et al., 2016; Ombelet & Johnson, 2016). Moreover, the dynamic interplay between sociocultural norms and religious beliefs is profoundly significant. In many parts of sub-Saharan Africa, surrogacy is often viewed through a biblical lens, leading to widespread resistance to the practice, which is frequently deemed religiously inappropriate (Ogar et al., 2018). Consequently, despite the potential advantages of surrogacy in assisting infertile couples to achieve parenthood, its acceptance is significantly limited by these deep-rooted cultural and religious convictions. The sociocultural and religious ramifications of surrogacy in Africa, combined with the existing regulatory hurdles, clearly demonstrate that aligning surrogacy practices with the cultural fabric of African societies is a complex endeavor. Nonetheless, this complexity should not impede the proactive efforts needed to develop efficient and ethical surrogacy frameworks across the continent.

Indeed, the urgent need for comprehensive surrogacy guidelines and regulations across African nations cannot be overstated. These regulations should not only ensure the protection of surrogate mothers and intended parents but also safeguard the rights of the children born from these arrangements. Crucially, legislative measures should be paired with robust societal education initiatives that aim to demystify surrogacy and address the stigmas associated with it (Umeora et al., 2014). Despite the myriad ethical, sociocultural, and regulatory challenges present in the surrogacy landscape across Africa, viable solutions are attainable. The experiences and advancements made by countries such as South Africa offer invaluable insights that can inform the development and implementation of comprehensive surrogacy policies in other African nations. However, realizing these objectives requires a concerted effort from various stakeholders, including policymakers, healthcare providers, religious and community leaders, and the general public. Through collaborative efforts, these groups can create an environment that not only supports the ethical practice of surrogacy but also respects cultural and religious norms, ultimately prioritizing the health and well-being of all parties involved (Adamu, 2020; Adelakun, 2018; Alabi, 2021).

### 3.2 Ethical Concerns of Surrogacy

Commercial surrogacy, which has become a prevalent contemporary fertility treatment, typically serves prospective parents with significant financial resources, often from developed countries. While this arrangement provides these individuals with the opportunity to become parents, it also introduces multiple health, emotional, and psychosocial risks to the surrogate mothers involved. This inherently imbalanced arrangement frequently involves less affluent and less educated women bearing children for wealthier clients, raising substantial ethical concerns. From a health perspective, surrogacy may expose both the surrogate and the developing fetus to various hazards (Afolabi, 2018; Ten Have, 2016). For example, to enhance the chances of pregnancy, surrogates are sometimes implanted with multiple embryos. This practice, which is widely regarded as ethically questionable, often puts commercial interests ahead of human health and well-being. Moreover, surrogacy carries emotional and psychosocial risks. Surrogates may develop a strong emotional attachment to the child they carry, potentially leading to significant stress and psychological challenges postpartum (Ogar et al., 2018). Another ethical issue arises from the practice of selective or nonmedical abortion, which occurs when fetuses do not meet the commissioning parents' expectations or exhibit congenital or genetic abnormalities. This practice, which can be seen as eugenic, devalues prenatal human life and raises serious bioethical questions. In some African nations like South Africa and Kenya, surrogacy is legally restricted to assisting infertile women. However, observations in Nigeria and Ghana indicate that fertility clinics may offer surrogacy services to healthy individuals seeking to avoid the physical conditions of pregnancy for various nonmedical reasons (Adelakun, 2018; Makinde et al., 2017, p. 20; Oti-Boadi & Oppong Asante, 2017). Such practices suggest that surrogacy is being used for eugenic or perfectionist purposes, a trend particularly noted in Ghana (Gerrits, 2018). This exploitation of surrogacy technology highlights the need for stringent ethical guidelines and rigorous regulatory oversight to protect all parties involved, particularly the surrogates, and to ensure that the practice of surrogacy upholds the dignity and value of human life.

### 3.3 Nature of the Immoral Act in Surrogacy

Surrogacy, which is rarely offered without charge, encompasses a range of commercial and personal interests that often compromise the health and well-being of the surrogates and unborn children. These prevailing interests render surrogacy inherently immoral as it commodifies human life, particularly vulnerable prenatal life, reducing it to a mere means to a material end. This commodification may lead to the loss of intrinsic value for human life and health, as secondary interests are prioritized over the fundamental well-being of vulnerable groups. The practice may also encourage selective and nonmedical abortions, which could foster eugenic tendencies and perfectionist interests in childbirth. Additionally, surrogacy has been linked to fertility tourism, the commodification of women's bodies, and a commercial approach to addressing fertility issues, as demonstrated by the "baby factory" cases in Nigeria and other illegal fertility market operations (Adelakun, 2018; Makinde et al., 2016, 2017, p. 20). It also facilitates the exploitation of poor and less educated women in regions with inadequate or nonexistent fertility laws by wealthier individuals from countries with stringent surrogacy regulations (Parry, 2018; Patel et al., 2018; Piersanti et al., 2021). Additionally, fertility clinics often adopt a paternalistic and medicalized approach to fertility issues, potentially interfering with natural biological processes and crossing ethical boundaries. Ultimately, surrogacy poses challenges to both state and global governance (Ten Have, 2016), underscoring the need for robust policies to curb potential moral decline and exploitation.

## *3.4 Ethical Principles and Values at Stake in Surrogacy*

This section critically examines the ethical and moral dilemmas associated with commercial surrogacy as an artificial fertility solution, analyzing various aspects of this practice through the perspectives of African scholars.

### 3.4.1 Justice, Fairness, and Equity

The principle of justice is crucial in the context of surrogacy as a fertility treatment. Surrogacy arrangements often exhibit an unequal power dynamic where the affluent benefit disproportionately compared to the underprivileged. Contracts typically favor wealthy commissioning parents, granting them exemptions from liability in cases where the surrogate suffers injuries or health complications during childbirth. This exception is distinct from the situation in South Africa, where courts uniquely oversee and arbitrate surrogacy agreements, ensuring a balanced approach to justice (ISF, n.d.). Such disparities highlight systemic failures in both global and local justice systems, intensifying the vulnerabilities of the poor and less educated. These practices call into question the fairness of government policies on surrogacy, which often appear skewed in favor of the affluent (Adamu, 2020; Ogar et al., 2018). In practice, achieving equity in surrogacy requires that surrogates, intended parents, and the resulting children receive fair and respectful treatment throughout the process. This should include transparent agreements, equitable compensation, and comprehensive support—medical, legal, and emotional—readily accessible to all parties involved.

### 3.4.2 Autonomy

While the principle of autonomy may initially appear less compromised in surrogacy arrangements, closer scrutiny reveals significant ethical breaches. Surrogates frequently report trauma upon relinquishing the baby, suggesting that their initial consent may not have fully encompassed an understanding of the potential consequences, health risks, emotional, and other psychosocial implications involved (Adamu, 2020; Dattijo et al., 2016). Moreover, socioeconomic pressures often compel women to become surrogates, highlighting the necessity for thorough scrutiny and transparency of the information provided to them. Additionally, the ethics of prenatal autonomy are compromised when embryos are discarded or terminated based on undesired traits or congenital defects (Gerrits, 2018; Oti-Boadi & Oppong Asante, 2017). This raises critical questions about the extent to which commissioning parents should have the authority to determine the fate of vulnerable prenatal lives.

### 3.4.3 Beneficence and Nonmaleficence

While surrogacy can offer substantial benefits to the parties involved, these often overshadow the potential risks. For example, surrogates and the embryos they carry are subjected to risks from multiple implantations, which can lead to long-term health issues. Moreover, the practice of selectively terminating fetuses and embryos that display "undesirable" traits highlights significant ethical concerns regarding nonmaleficence. The rise of illegal surrogacy operations and baby factories in Africa (Makinde et al., 2016, 2017; Rashid, 2018), as well as the growing prevalence of transnational surrogacy in Ghana (Gerrits, 2018), prompt a critical examination of whose interests are truly being served. There is a growing concern that commercial interests are often prioritized over the well-being of the surrogates and, potentially, the children they bear (Adamu, 2020; Ogar et al., 2018).

### 3.4.4 Communitarianism (Including Family and Cultural-Religious Values)

From a communitarian ethical perspective, the practice of surrogacy in Africa is perceived as indicative of a societal moral decline concerning fertility (Alabi, 2021; Ogar et al., 2018; Pheme et al., 2020; Umeora et al., 2014). This acceptance within society often reflects diminishing respect for traditional childbirth methods. Moreover, the growing sociocultural endorsement of surrogacy, exemplified by the South African scenario where it is regarded as a noble aid for infertile couples seeking parenthood (ISF, n.d.), provokes ethical concerns. In many parts of Africa, the decision to engage in surrogacy extends beyond personal choice, influenced heavily by family dynamics and societal pressures, particularly in patriarchal societies. Many African scholars, including Ogar and his colleagues, argue that this trend signifies a deterioration in communal and religious values, highlighting that there are more dignified responses to infertility, such as adoption (Ogar et al., 2018). Furthermore, the divergent and inconsistent views among various societal stakeholders suggest a deepening crisis in the societal perception of the morality, ethics, and legality of surrogacy (Umeora et al., 2014).

### 3.4.5 Virtue and Natural Law Ethics

Surrogacy poses significant challenges to virtue ethics and natural law, as it involves commodification of women's bodies and interference with natural childbirth processes. On one hand, physicians are seen as overstepping their bounds by medicalizing fertility issues. On the other hand, surrogates and their families are often seen as morally deficient for choosing surrogacy as a socioeconomic empowerment strategy, despite the associated health risks (Alabi, 2021; Ogar et al., 2018; Pheme et al., 2020). In many situations, fertility problems are perceived as a test of the ability for perseverance on the part of couples seeking childbirth. As Ogar and colleagues argue:

> …appeal to the Christian Bible reveals in many chapters and books of the Bible that it is God's single power or prerogative to give or not give children. All attempts past and present in man contravening the will of God had only ended him (man) in woes, sorrow and pain… (Ogar et al., 2018)

Furthermore, surrogacy largely interferes with a natural process of childbirth and hence a breach of natural law ethics on the parts of all actors involved in the process, including the larger society that perceives infertility as an object of stigma. Finally, the lack of regulations or legal frameworks in most African countries fails to deter acts of criminality and exploitation in surrogacy.

### 3.4.6 Personalism (Value for the Sanctity of Human Life and Respect for Human Dignity)

Personalism ethics, emphasizing the sanctity of life and human dignity, is a crucial aspect of the ethical discourse surrounding surrogacy within the African context. The commodification of surrogacy places material and socioeconomic interests above the intrinsic value of life and the dignity of human beings at all stages of life. The increasing medicalization of fertility concerns further erodes the dignity of human beings, threatening the self-identity of both surrogates and the children they bear, thereby reducing individuals to mere instruments for achieving material ends. This situation is further complicated as fertile women may opt to avoid pregnancy due to career or cosmetic reasons. Additionally, the rise of surrogacy driven by eugenic or perfectionist motives, where couples pursue their own material and self-interests rather than valuing the dignity and sanctity of the lives of surrogates and their children, presents significant ethical challenges. These practices call for a reevaluation of values to ensure that respect for human dignity and life remains paramount in surrogacy arrangements.

### 3.4.7 Vulnerability: Protection of Persons

Surrogacy is often a practice involving imbalanced negotiation of benefits. The surrogates, especially those from poor backgrounds, are particularly vulnerable to exploitation in surrogacy arrangements. This vulnerability is not unlike that faced by workers in other industries where power imbalances are stark, such as migrant labor or the gig economy. In these contexts, individuals often accept conditions and compensations that may not fully compensate them for their services and risks due to economic necessity. In surrogacy, this exploitation could manifest in inadequate compensation, poor health care, and coerced consent, where surrogates may not be fully aware of the risks or their rights. The ethical challenge is to create a framework where surrogates' rights are protected from being exploited by not only the system but also the commissioning parent(s), as well as the fertility clinics and the interests of middle men.

### 3.4.8 Sustainability: Building Long-Term Ethical Practices

The principle of sustainability in surrogacy refers to the development of practices that are not only effective in the short term but also ethical and beneficial in the long run for all parties, including the children born out of surrogacy arrangements. This concept can be compared to environmental sustainability, which seeks to meet present needs without compromising the ability of future generations to meet their needs. In the practice of surrogacy, sustainable practices involve ensuring that surrogacy is conducted in a way that maintains the long-term health and psychological well-being of surrogates and children. This means careful consideration of the medical, emotional, and social impacts of surrogacy and implementing practices that minimize discrimination of children born of surrogates, as well as for surrogates, such as ensuring ongoing psychological support and monitoring the long-term health of surrogates.

### 3.4.9 Access: Expanding Safe and Ethical Surrogacy Services

Improving access to safe and ethical surrogacy services involves addressing disparities in availability and quality of surrogacy services between different regions and socioeconomic groups. This challenge is similar to that faced in global health (Afolabi, 2018; Ten Have, 2016), where there is a need to ensure that life-saving medical interventions are available not just in wealthy countries but globally. Ensuring broad access to surrogacy means implementing policies that make ethical surrogacy practices widely available and affordable, and not just an option for the wealthy or those in developed nations. It also involves educating potential surrogates and intended parents about their rights and the ethical dimensions of surrogacy to foster informed decision-making.

### 3.5 *Implications for Health and Life in Surrogacy: A Proposed Way Forward*

Surrogacy carries significant implications for the health and well-being of both surrogates and the children born from such arrangements. African scholars have highlighted the considerable physical, psychological, and social harms associated with surrogacy, both in the short term and long term (Adamu, 2020; Alabi, 2021; Ogar et al., 2018; Oti-Boadi & Oppong Asante, 2017; Umeora et al., 2014). Challenges such as multiple embryo implantations can lead to gestational complications akin to those seen in traditional pregnancies, potentially jeopardizing the health of resultant children by increasing the likelihood of weaker embryos being outcompeted. Ethical concerns also arise regarding the fate of embryos with "undesired traits," which are often terminated, and the surplus embryos that may be destroyed after pregnancy. Additionally, egg donors, particularly those involved in genetic surrogacy, face risks from the long-term health effects of hormone injections used in in vitro fertilization. Moreover, children born through surrogacy may confront identity issues, as seen in the case of a surrogate-born child to British citizens in Kenya who was denied UK citizenship (Rashid, 2018). These issues emphasize the urgent need for robust transborder surrogacy regulation and international dialogue (Adelakun, 2018; Alabi, 2021; López Guzmán & Aparisi Miralles, 2012; Ogar et al., 2018; Piersanti et al., 2021; Rashid, 2018; Söderström-Anttila et al., 2016; Ten Have, 2016). In response, the review in this chapter advocates for a three-pronged strategy to address these challenges within Africa:

1. *Robust Regulation*: Implement stringent, enforceable guidelines and regulations at both national and international levels to curb surrogacy-related transborder criminal activities. While achieving consensus may be difficult due to ethical, moral, and cultural differences, it is essential to promote global discussions on surrogacy. Models like South Africa's, which require judicial oversight for all surrogacy arrangements, can offer a framework for protecting all parties involved, especially the vulnerable (Ogar et al., 2018).
2. *Community Sensitization and Advocacy*: In Africa's communitarian society, it is vital to engage in community education and advocacy to change public perceptions and reduce stigmas associated with infertility. This approach can help decrease the demand for potentially unethical surrogacy practices.
3. *Ethical Education and Awareness*: Promote ethical education to foster a more nuanced understanding of infertility issues within the framework of religious and natural law principles, emphasizing the respect for human dignity at all life stages. The aim is to ensure that surrogacy is conducted in a manner that respects the dignity and self-determination of surrogates and safeguards the rights and welfare of the children involved.

By pursuing these strategies, it is possible to develop actionable solutions that uphold ethically sound practices in surrogacy, ensuring it contributes positively to societal welfare and respects the fundamental rights and health of all individuals involved.

## 4 Conclusion

This chapter critically examines the practice of surrogacy as a fertility solution within an African context, drawing insights from the perspectives of African scholars. Surrogacy, whether procured locally or through transborder transactions, and whether driven by commercial or altruistic intents, presents two primary categories of issues. First, ethical dilemmas are prominent, including concerns around embryonic handling, the devaluation of human life at the prenatal stage, the implications of "playing God," and the medicalization of infertility. Second, broader moral concerns emerge from the commodification of women's bodies, the exploitation of vulnerable women, especially those who are impoverished and illiterate, the commercialization of babies, and the complex issues surrounding the rights and identities—both personal and national—of children born from surrogate arrangements. The extent of these problems challenges various global bioethical principles and values, significantly infringing upon the tenets of personalism ethics. This ethical framework emphasizes respect for the ontological nature of human beings and the sanctity of life at all stages, prioritizing these above material, socioeconomic, and cultural-religious interests. To address these issues, this chapter proposes a multifaceted approach not aimed at banning surrogacy but at regulating the practice alongside other artificial fertility solutions to mitigate the risks and adverse impacts on the health and well-being of women and children involved. The recommendations put forth are threefold: (1) **Legislation and Regulation**: Establish robust legal frameworks to govern surrogacy practices, ensuring they protect the rights and welfare of all parties involved, especially the surrogates and the children born from these arrangements. (2) **Cultural-Religious Integrity**: Promote the alignment of surrogacy practices with prevailing cultural and religious values, enhancing acceptance and reducing conflicts that arise from ethical and moral discrepancies. (3) **Ethical Literacy and Awareness**: Advance ethical literacy and raise awareness across the broader society to ensure a well-informed public that can engage in thoughtful deliberation on the complexities of surrogacy and other fertility solutions. By implementing these recommendations, the aim is to foster a more ethical, respectful, and humane approach to surrogacy in Africa, ensuring not only that the practice contributes positively to society and respects the fundamental dignity and rights of all individuals involved but also to protect vulnerable persons and ensure access to fertility solutions in an equitable manner.

## References

Adamu, C. O. (2020). Ethical issues in commercial gestational surrogacy. *AMAMIHE: Journal of Applied Philosophy, 18*(4), 88–117. https://doi.org/10.13140/RG.2.2.26043.64803

Adelakun, O. S. (2018). The concept of surrogacy in Nigeria: Issues, prospects and challenges. *African Human Rights Law Journal, 18*(2), 605–625. https://doi.org/10.17159/1996-2096/2018/v18n2a8

Afolabi, M. O. (2018). *Public health disasters: A global ethical framework* (Vol. 12). Springer. https://doi.org/10.1007/978-3-319-92765-7

Alabi, O. J. (2021). Perceptions of surrogacy within the Yoruba socio-cultural context of Ado-Ekiti, Nigeria. *F1000Research*, 9. https://doi.org/10.12688/f1000research.20999.3

Dattijo, L., Andreadis, N., Aminu, B., Umar, N., & Black, K. (2016). Knowledge of infertility among infertile women in Bauchi, northern Nigeria. *International Journal of Women's Health and Reproduction Sciences, 4*(3), 103–109. https://doi.org/10.15296/ijwhr.2016.25

Dawes, A. (2009). The South African children's act. *Journal of Child & Adolescent Mental Health, 21*(2), iii–vi. Taylor & Francis.

Gerrits, T. (2018). Reproductive travel to Ghana: Testimonies, transnational relationships, and stratified reproduction. *Medical Anthropology: Cross Cultural Studies in Health and Illness, 37*(2), 131–144. https://doi.org/10.1080/01459740.2017.1419223

ICCA. (2019). *Africa legal review*. ICCA.

ISF. (n.d.). *SOUTH AFRICA | International surrogacy forum*. Retrieved June 20, 2021, from http://www.internationalsurrogacyforum.com/content/south-africa

López Guzmán, J., & Aparisi Miralles, A. (2012). An approach to the legal and ethical problem of surrogate motherhood. *Cuadernos de bioética: Revista oficial de la Asociación Española de Bioética y Ética Médica, 23*(78), 253–267.

Makinde, O. A., Makinde, O. O., Olaleye, O., Brown, B., & Odimegwu, C. O. (2016). Baby factories taint surrogacy in Nigeria. *Reproductive Biomedicine Online, 32*(1), 6–8. https://doi.org/10.1016/j.rbmo.2015.10.001

Makinde, O. A., Olaleye, O., Makinde, O. O., Huntley, S. S., & Brown, B. (2017). Baby factories in Nigeria: Starting the discussion toward a National Prevention Policy. *Trauma, Violence, and Abuse, 18*(1), 98–105. https://doi.org/10.1177/1524838015591588

Ogar, J. N., Leonard, N., & Akpan Bassey, S. (2018). Issues related to infertility in Africa: An ethical scan. *International Journal of Biomedical and Advance Research, 9*(9). https://doi.org/10.7439/ijbar

Ombelet, W., & Johnson, M. (2016). Baby factories in Nigeria: A new and challenging source of abuse. *Reproductive Biomedicine Online, 32*(1), 4–5.

Oti-Boadi, M., & Oppong Asante, K. (2017). Psychological health and religious coping of Ghanaian women with infertility. *BioPsychoSocial Medicine, 11*(1). https://doi.org/10.1186/s13030-017-0105-9

Parry, B. (2018). Surrogate labour: Exceptional for whom? *Economy and Society, 47*(2), 214–233. https://doi.org/10.1080/03085147.2018.1487180

Patel, N. H., Jadeja, Y. D., Bhadarka, H. K., Patel, M. N., Patel, N. H., & Sodagar, N. R. (2018). Insight into different aspects of surrogacy practices. *Journal of Human Reproductive Sciences, 11*(3), 212.

Pheme, M. J., Sithole, S. L., & Malema, R. N. (2020). The perceptions of childless women on surrogacy as an assisted reproductive technique (art) in Lebowakgomo, Capricorn District, Limpopo Province. *Social Work (South Africa), 56*(3), 326–346. https://doi.org/10.15270/56-4-861

Piersanti, V., Consalvo, F., Signore, F., Del Rio, A., & Zaami, S. (2021). Surrogacy and "procreative tourism". What does the future hold from the ethical and legal perspectives? *Medicina (Lithuania), 57*(1), 1–16. https://doi.org/10.3390/medicina57010047. MDPI AG.

Rashid, F. (2018). *The legal and regulatory framework of surrogacy theory and practice* (Issue December). University of Nairobi.

Söderström-Anttila, V., Wennerholm, U. B., Loft, A., Pinborg, A., Aittomäki, K., Romundstad, L. B., & Bergh, C. (2016). Surrogacy: Outcomes for surrogate mothers, children and the resulting families – A systematic review. *Human Reproduction Update, 22*(2), 260–276. https://doi.org/10.1093/humupd/dmv046. Oxford University Press.

Ten Have, H. (2016). *Global bioethics: An introduction*. Routledge. https://www.taylorfrancis.com/books/mono/10.4324/9781315648378/global-bioethics-henk-ten

Umeora, O. U., Umeora, M., Chukwuneke, F., & Emma-Echiegu, N. (2014). Surrogacy in Nigeria: Legal, ethical, socio cultural, psychological and religious musings. *African Journal of Medical and Health Sciences, 13*(2), 105. https://doi.org/10.4103/2384-5589.144579

# Chapter 9
# Ethical and Legal Issues on Embryo Adoption, Donation, and Surrogate Motherhood: A Kenyan Perspective

**Anthony Kebira Nyamache**

**Abstract** Most countries, including Kenya, have not yet developed a legal or policy framework to govern assisted reproductive technologies since its inception in 1981. Even though surrogacy, in which a woman carries a pregnancy for another or infertile woman and accepts donated embryos from another couple to conceive and give birth to an adopted child, could be beneficial in addressing infertility in couples, this technology has no law to regulate it. However, assisted reproductive technologies; surrogacy, embryo adoption or donation, ethically accepted means to treat infertility, these technologies have multitude of ethical and legal issues appealing for an address.

This study was therefore aimed at identifying the main ethical and legal issues arising from embryo donation and surrogacy motherhood in treating infertility from a Kenyan perspective. A qualitative interdisciplinary nonsystematic literature review on ethical and legal analysis of existing and pending policy, laws, and ethical issues related to donated gametes, adopted embryos, and surrogacy motherhood was conducted. Mendeley, Google Scholar, and PubMed databases were searched in the English language for the themes "Ethical and legal issues in assisted reproductive technology AND Kenya," "Ethical and legal issues of embryo donation AND Kenya," and "Ethical and legal issues of surrogacy, AND Kenya." Using the above themes or phrases, a total of 82 articles were retrieved with only 4 being those that dealt specifically with subject matter analyzed. Literature reviews on ethical and legal reviews were analyzed and core ethical and legal issues were identified.

The study revealed that the practice of surrogacy donated embryos or adoption in Kenya operates under the Law of Contract Act and there is no law or policy to regulate the practice. Commercial surrogacy, the treatment of children as commodities,

A. K. Nyamache (✉)
Department of Biochemistry, Microbiology and Biotechnology, Kenyatta University, Nairobi, Kenya
e-mail: nyamache.anthony@ku.ac.ke

 
L. Omutoko, W. Jaoko (eds.), *Bioethics from the Global South*, Advancing Global Bioethics 21, https://doi.org/10.1007/978-3-031-77669-4_9

and exploitation of women are issues that rose. Other pertinent legal issues included parental rights and the effects of manipulating embryos. However, some of the ethical issues raised by this technology include the lack of informed permission, human embryo tampering, physical and psychological trauma, threats to the traditional nuclear family's structure, and identity crises experienced by the children born out of this marriage.

This study demonstrates that the use of ART is an alternative for treating female infertility. Accessibility to this service has resulted into abuse and exploitation of women who might otherwise be at risk of not having access to it for infertility management. Therefore, in order to protect the rights of those affected and to solve the stated ethical issues, the 2022 assisted reproductive technology bill and policy must be expedited as quickly as possible.

**Keywords** Surrogacy · Legal · Ethics · Embryo donation · Assisted reproduction · Kenya

## 1 Background

Most couples around the world, especially in Africa, consider having or raising children as an important life event. Since the birth of a child is thought to satisfy everyone's need for procreation, it is also considered one's completeness in life (Roupa et al., 2009). Infertility affects millions of people with an estimated one in six people in their reproductive age worldwide experiencing infertility (WHO, 2020). It is thus a stressful experience in these couples when they find it difficult to become pregnant even after engaging in frequent, unprotected sexual activity for a maximum of 12 months (Njogu et al., 2022). The most common causes of male infertility are azoospermia, or aberrant sperm shape and motility, low sperm count, or difficulties with semen ejaculation. In females, many anomalies of the ovaries, fallopian tubes, uterus, and endocrine systems may cause infertility (Gore et al., 2015; Segal & Giudice, 2019; Zegers-Hochschild et al., 2013). In addition, venereal infections and other conditions like fibroids are among the predisposed risks to infertility. Therefore, all people, regardless of their gender, ethnicity, nationality, race, religion, or socioeconomic level, are affected by infertility (Njogu et al., 2022). Infertility issues affect between 8% and 12% of couples globally, with most of them residing in developing countries (WHO, 2020). Among these, about 50% often seek medical intervention for a child from a wide range of options including assisted reproduction technologies (ARTs) (Njogu et al., 2022).

Infertility is a global major public health problem facing many couples that affects not only their social life but their marriage (Ombelet, 2011). It has therefore remained a big challenge for many families trying to get children of their own. As a public health issue, most countries are affected. Issues of infertility have therefore posed serious social, psychological, financial, legal, and ethical implications in society (Shreffler et al., 2010). The witnessed infertility-associated social stigma to couples for being childless have made them feel abandoned especially those residing in developing countries (Tabong & Adongo, 2013). Even if these couple could opt to remain without children, they will still face stigma, ridicule, family pressures, and psychological torture. Therefore, these couples will more often find themselves in a dilemma of either using these ART methods to get a child or remaining without a child, a situation worth studying.

Use of ART, embryo adoption, and surrogacy are ethically widely accepted for the treatment of infertility, but their cost is so high beyond reach. The vulnerability of Kenyan community not accessing ART is exacerbated by the current health inequities. Thus, the difficulty in reaching the poor has increased due to the disparities in access to healthcare and economic resources. It is estimated that nine of ten countries with highest infertility rates are found in Africa makes it more urgent for medical intervention. Though the prevalence of infertility in Kenya is still unclear, the estimated 26.1% reproductive consultations associated with delayed fertility with 50% attributed to tubal blockages and 15% to male characteristics indicate that the cases could be high what is indicated here (Murage et al., 2011).

With high-cost ART, more often infertile couples have also tried to first seek alternative treatments like surgery, treatment of underling congenital issues, or lifestyle modification before resoughting for inaccessible ART (Njogu et al., 2022). If the accessibility to ART services could be increased, probably the infertility challenge could be half solved. Moreover, if the law could be enacted to regulate cost ART, probably its accessibility could be improved to reduce the number of couples without children. Treatment for infertility has remained with the use of ART: either surrogacy, embryo donation, or adoption. The use of embryo adoption, donation, and surrogacy motherhood as a subject matter has become the problem rather than the research questions on ART options (Tabong & Adongo, 2013). Adaptation of these methods has attracted a multitude of ethical, psychological, and legal issues. Despite the introduction of ART as late as 1978, most countries have not yet initiated regulation of ART (Tabong & Adongo, 2013). Kenya, like most countries, has not yet enacted any law to regulate ART. The lack of a law to control the cost or regulate ART has led to the exploitation of vulnerable infertile couples (Tabong & Adongo, 2013). In addition, surrogates have been exploited.

The ART treatment involves in vitro handling of human oocytes and sperm to achieve pregnancy with the expectation of artificial insemination. The technology involves fertilization of the women's eggs and men's sperm in vitro to form embryos (Zegers-Hochschild et al., 2009) with the formed embryos transferred into a woman's uterus for growth to a full-term baby. The use of surrogacy, embryo adoption, or donation is among assisted reproductive technology (ART) options that are often used in infertility treatment. In surrogacy, a woman is appointed to act in the place of another and it can be partial with the surrogate's egg being fertilized with the sperm of the intended father (Gunnarsson Payne et al., 2020) or full, where the eggs of the intended mother and father or donor are used with the baby having no genetic connection with the surrogate mother (Gunnarsson Payne et al., 2020). Where there is a third-party reproductive practice, like in surrogacy, consent and a contract have to be sought from the involved parties (Saxena et al., 2012). In a surrogacy, the arrangement could be commercial with the surrogate being paid or altruistic with no compensation for intended parents like medical bills or direct costs that could be associated with the pregnancy (Brandão & Garrido, 2023). Between the two surrogacy arrangements, partial surrogacy was more popular in earlier years, but in the recent past commercial surrogacy has become the most common (Ragoné, 1996; Gunnarsson Payne et al., 2020).

Ethical and legal issues arise from the process of embryo donation. During egg "harvesting" the biological mother does not always receive all the embryos obtained after fertilization, and the remaining embryos are often donated to needy infertile couples after consent (Patterson, 2015). Embryo donation has been practiced by many and its demand has continued to increase (Wallach & Robertson, 1995). Notably, most couples who are unable to produce eggs and sperms often opt for embryo donation from anonymous donors. Despite the increasing demand for ART services, there is currently no legal framework to regulate surrogacy, embryo donation, or embryo adoption; however, there are arrangements to enact a law. The operations of ART are at the moment being operated under the Law of Contract Act. This contract only outlines the roles, responsibilities, and expectations of all involved parties as a way to prevent any future dispute. Other countries have also initiated legislative laws to regulate ART in some European countries (Präg & Mills, 2017). Some jurisdictions such as the United Kingdom and some states of Australia have elaborated regulations with only United States relying on professional guidelines and general medical practice to govern ART (Adamson, 2005; Frith & Blyth, 2014). The regulation of ART therefore remains a contested area. With nonexistent regulatory ART laws to protect the rights of the surrogate mothers, intended mothers, or third parties in adopted embryos, certain delicate ethical and legal issues are often left posing threats to implementation of ART (Saxena et al., 2012). Failure to regulate ART and provide mechanisms to ensure the protection of involved parties is likely to expose populations to untold suffering. I argue that enactment of an ART law will regulate unfair practices and protect infertile couples from exploitation and high cost of services culminating in a reduction of legal battles and respect of individual rights. The implementation of surrogacy and embryo-adopted motherhood has attracted ethical, legal, and psychological issues that need attention. With the expansion and use of surrogacy and adopted embryos as ART options, there is a need to address related and legal issues that have become paramount.

## 2 Methodology

A descriptive and analytical approach was adopted in assessing the current status of ethical or legal issues arising for ART technologies in the country. References were drawn from electronic articles, periodicals, and thesis. After formulating a research problem with available data, literature reviews on ethical and legal associated with embryo adoption, donation and surrogacy motherhood were conducted from databases of PubMed and Google Scholar with an estimated 477 articles retried. The themes and ethical and legal issues on ARTs: "Ethical issues of embryo adoption AND Kenya," "Ethical issues of embryo donation AND Kenya," "Ethical issues of surrogacy AND Kenya," and "Ethical issues in assisted reproductive technology AND Kenya" were searched from these databases. Out of the 477 articles, 82 were retrieved and reviewed. The themes were filtered using search terms to confirm relevance, after which an analysis based on legal and ethical issues was conducted.

## 3 Ethical and Legal Concerns from a Kenyan Perspective

The practice of surrogacy, embryo adoption, and donation has raised complex moral, ethical, and legal questions throughout the world, including Kenya. The child, intended or commissioned parents, healthcare professionals, surrogate mother, sperm or egg donors, and the government must all be taken into account when applying bioethical principles to the decision-making process related to ART, surrogacy, embryo donation, and adoption. Several experts have conducted bioethical analysis and proposed solutions to this practice. This chapter addresses surrogacy, embryo adoption, and donation currently practiced in Kenya as a bioethical and legal dilemma. This dilemma shows conflict between alternatives if one chooses any of them, which could compromise ethical principles leading to violation of one kind or another. Initially, the choices made by individuals to solve infertility appear unsolved but there could be several reasons for supporting or a particular decision. For ethical acceptability, there must be scientific evidence which could be either debatable or conclusive (Rozée et al., 2020). Nevertheless, in such situations, decisions must be made by and for the society as a whole.

The American medical bioethicists Tom Beauchamp and James Childress developed the four prima-facie principles of bioethics to assist physicians, nurses, and other healthcare professionals in resolving the ethical dilemmas that they invariably encountered when providing treatment for their patients (Childress & Beauchamp, 1994). They also expanded upon the idea put forth by Oxford philosopher David Ross in the 1930s that ethical principles should be prima facie, or conditional, for a stronger or more compelling principle to prevail over a weaker one in a given circumstance. In addition, medical ethics choices should be made on individual basis by applying the prima facie principles to each circumstance (Riddle, 2022). The four principles of bioethics are nonmalfeasance—to cause no harm, beneficence—to effect a cure, autonomy—to respect patients' independence, and justice—to treat patients fairly and without discrimination. These guiding bioethical principles offer broad directions for addressing any bioethical issues that cross national, cultural, and religious boundaries.

## *3.1 Ethical Issues*

### 3.1.1 Nonmalfeasance

The principle of nonmalfeasance states that you should "cause no harm" which includes not to kill, not to cause pain, and not to incapacitate or deprive of goods (Childress & Beauchamp, 1994). Following the unregulated ART by law or policy in Kenya, altruistic and commercial surrogacy is in operation. However, in both cases, surrogates and children born from these procedures are subjected to physical and psychological harm. If the government could be addressing poverty, probably surrogates due would not engage in this risky procedure. In Kenya, commercial surrogacy is the most common. It can be argued that during the gestation period, surrogates are exposed to physical harm due to the complex medical procedure. The surrogate mother is not genetically related to the fetus, the pregnancy is artificial, and the preparation of the surrogate and the transfer of the embryo require ideal medical attention. Usually, the surrogate's natural ovulation cycle is suppressed and controlled using birth control pills and hormone injections. Then estrogen shots are administered to aid in the development of her uterine lining. After that progesterone injections are given every day after the embryo is transplanted until the surrogate's body eventually accepts the pregnancy and starts to support it. The surrogate has to experience the following adverse side effects due to the administered drugs: hot flashes, mood swings, headaches, bloating, spotting, uterine cramps, breast fullness, dizziness, and vaginal discomfort (Agnafors, 2014). All these experiences have to be endured despite the inflicted harm. Nevertheless, they are also at risk of multiple pregnancy and associated pain during delivery not forgetting the social stigma and ostracism one faces from relatives and the community (Brezina & Zhao, 2012; Hodson et al., 2019; Riddle, 2022).

For the egg or embryo donors, they are also at risk of ovarian hyperstimulation syndrome which could occur during the egg stimulation and egg harvesting that predisposes these donors to harm.

Technologies applied in embryo donation are faced with an ethical question, which is the extent of anonymity of the donors or surrogate mother preservation. Children born out of these technologies will often undergo social stigma and psychological harm upon realization of their own circumstances of their conception (Childress & Beauchamp, 1994; Jadva et al., 2003; Yee et al., 2020; Brandão & Garrido, 2023). Though anonymity is more preferred by most embryo donors or surrogates, this is often considered unethical (Alizadeh & Samani, 2014). Children born through these procedures face various challenges, for example, when they try to trace their ancestral heritage, they often get confused and feel isolated from the rest of the family members. The process of tracing ancestral or genetic mothers is often made difficult and challenging based on the signed agreement made with their rearing parents (Mitra, 2018; Montanari Vergallo et al., 2018). This can result in the children viewing themselves as commodities and transactions, hence finding themselves in emotional distress. Upon discovery of their origins, these children are likely to feel that their integrity and rights have been violated. On the other hand, the surrogate mothers also feel exploited, violated, and disrespected. Despite the assisted reproductive technologies being a good option for the treatment of infertility, their use could put one at risk of giving birth to a child with genetic disorders or defects despite its cost (Namavar Jahromi et al., 2018). The intended parents with a child having a genetic disorder could be subjected to psychological trauma, stress, and stigma. In some instances, some of intended mothers could find themselves undergoing stressful moments when they cannot control or direct the surrogates as per their wish during gestation. The situation could also be worsened if the distance between the surrogate mother and intended parents is too far between. Surrogacy is a risky undertaking because one is expected to experience physical and physiological harm during delivery or even end up dying from complications (Jadva et al., 2003; Yee et al., 2020). Both the intended parents and the surrogates are also faced with high levels of social stigma and ostracism. To avoid this, involved parties prefer to do this in a secure or secretive place away from relatives or people who know them.

### 3.1.2 Beneficence

The principle of beneficence states that physicians must promote the well-being of others, which is interpreted to mean acting for the benefit of the patient. They must defend the rights of others, prevent harm, or remove conditions that could cause harm or rescue one from danger (VarKey, 2021). According to this principle, let us consider the surrogate mother under treatment as patient while commissioning parents as the client. It would appear that the clients who are the intended or commissioned parents benefit more from surrogacy by getting a child according to beneficence (Mitra, 2018). Equally, the surrogate mother will also benefit by meeting her satisfaction of aiding others in getting a child and considerable compensation. Similarly the offspring born from this process will get a right to live (Brinsden, 2003; Brandão & Garrido, 2023). Considering embryo donation and embryo adoption technologies, the infertile couple seeking infertility treatment will benefit by getting a child. Embryo donation and embryo adoption are both ethically accepted practices that are beneficial to infertile couples. Embryo donation does not present any risks or problems beyond those presented by embryo adoption or surrogacy. This practice is arguably more respectful of embryos than IVF by giving them another opportunity to be implanted that might otherwise be discarded (Wallach & Robertson, 1995). What is unique with embryo donation is that a child born to a couple has no parental genetic link to it. This makes it different from other forms of

ART where one or both rearing parents are genetic parents to the child. For instance, in gestation surrogacy, both rearing parents are genetic parents even with a genetic mother not carrying the pregnancy or full surrogacy, with the rearing mother having no genetic link except the father. Children born of embryo donation technology appear less likely to suffer from social or psychological harm compared to those born from surrogate, embryo adoption, or postnatal adoption. From the child's perspective, the embryo donation would be more preferable to postnatal adoption since the rearing mother will have carried the pregnancy from a committed father. However, the surrogate mother as a patient is not yet protected. The surrogate mother would consider herself as used or a commercial product; hence her dignity is violated. Participation in surrogacy also exposes them to the risk of procedure which could cause physical and psychological harm in exchange for a fee. It would therefore be prudent to reevaluate the benefit of the surrogate mother and see how her benefits would be promoted throughout the surrogacy process. Ethically, it is expected that for one to gain from surrogate motherhood, benefits should outweigh harm which can either be physical, psychological, emotional, or social (Montanari Vergallo et al., 2018). It has been observed that in commercial surrogacy, the surrogate mothers engage in surrogacy due to hard economic pressure to cater for their families (Jadva, 2020). They consider financial gain as a positive benefit. But in altruistic surrogacy, the patient or surrogate mother would consider this as a loss. Despite nonexistent laws or policies to regulate ART, surrogacy has gained popularity in Kenya since 2007 with the demand increasing within the country and across borders.

### 3.1.3 Autonomy

This is the right of an individual to make their own decision or self-determination excluding external interference (Childress & Beauchamp, 1994). According to ART, this is a "right" and has diverse interpretations. In medical field, autonomy is defined as "self-determination" free from external interference from other individuals or personal weakness to make a choice/decision like inadequate understanding or faulty reasoning (Mackenzie et al., 2020). For the uptake of any ART options by the infertile couples, they should be able to choose freely what they want, unless their actions could cause serious harm to others or limit others liberty. We consider ART as practiced in Kenya whether its autonomy is actually upheld or preserved. If surrogacy, embryo donation, or adoption is performed by free will by all participants, intended parents, the surrogate, or the gametes donors, then one would assume that autonomy is guaranteed (Smith et al., 2013). It will be also be considered, a violation of autonomy and free will if ART, surrogacy, embryo donor, or embryo adoption were to be outlawed. This would prevent intended parents and the couples who are adopting, donating, or surrogating from having the choice to engage in it (Gracias, 2021). Considering the offspring from this process, this child would not be autonomous to make a decision and hence the restricted right to do so (Sifris, 2015; Wade, 2020).

It is observed that although ART would help to create one family life, it could also potentially destroy another (Ekberg, 2014). For instance, the surrogate mother after the 9 months of a job is over, one would be expected back to their marriage in a society riddled with stigma. It can also be noted that that freedom to contract could be intended to be the autonomy of the surrogate, but in reality, this could be constrained. In drawing the contract in commercial surrogacy, it more often infringes upon their autonomy. The surrogate's behaviors and emotions are restricted, and they have a voice in the entire progress. Nevertheless, the contract mostly favors the intended parents and the health providers. It should be noted that most surrogate mothers who engage in commercial surrogacy are mostly those from poor backgrounds, and therefore poverty and desperation could drive their motive to participate (Rozée et al., 2020). This raises the question of whether the surrogate decision to participate in the procedure in the first place was made voluntarily. Some ART clinics have been established as a result of the increasing demand for surrogacy services. Medical professionals, nurses, and clinic employees who profit financially from surrogacy regularly visit underprivileged areas and universities to find young women willing to serve as surrogates (Wade, 2017). Most of these surrogates frequently favor commercial surrogacy contracts because of familial obligations and poverty. During recruitment, the surrogate has to forgo her autonomy and be screened and approved or denied. If recruited, you have no choices on who the intended parents will be. Similar to a closed adoption, you are also not allowed to meet the prospective parents (Sharma, 2014). During delivery, the newborn is handed over to the intended parents or commissioned parents without the surrogate's autonomous and informed consent. The contract does not allow the surrogate time to change her mind or even object handing over of the infant. Nevertheless, it is assumed that these surrogates may not have the capacity to contest the terms of the surrogate contract.

### 3.1.4 Justice

Justice is defined as the concept of fairness and equality in the distribution of scarce health resources and the decision of who will receive what treatment. Justice is therefore a principle with four components: distributive justice, respect for the law, rights, and retributive justice. Distributive justice is concerned with the equitable allocation of resources; the second refers to whether the fact that an act is or is not against the law, is of moral relevance; rights are considered to be special advantages with correlative duties to provide them; retributive justice refers to making right when a wrong has been perpetrated. If all people would have an equal access to ART, the principle justice would be guaranteed. However, inequitable distribution of access to ART care has remained the most obvious ethical challenge. Most countries have economic barriers to ART, which prevents most economically disadvantaged couples from accessing it (Chambers et al., 2009). The use of embryo adoption, donation, or surrogacy services by most countries including Kenya to treat infertility could be considered as justice served.

The provision of ART services has attracted several ethical concerns. It is expected that for justice to be served, all burdens and risks should be spread equally to all parties involved (Pennings, 2011; Montanari Vergallo, 2019). To access the ART services in Kenya is, however, expensive and only limited to the rich or able society. The limited access to these services is unethical and shows an inequitable access to important public services. Despite the cost of ART being high, it varies across the providers in the country but is still exorbitantly high. With an increased demand for ART services even from the neighboring countries, the cost of ART services has been rising leaving the impoverished and vulnerable poor without access. The lack of government to provide ART services leaves Kenyans to be exploited by private sectors. Lack of regulation of ARTs services results into unstandardized cost and denies access. The created disparities or inequity of these services confirms unethical service provision. Due to the high cost of ART services, access to healthcare and the right to parent are frequently restricted. Those who view surrogacy as social injustice often argue that surrogacy exploits women by commodifying their wombs to meet the reproductive needs of the disadvantaged in the community (Kindo et al., 2020). Even though surrogacy is known for its treatment of infertility, it is in demand even among unrecognized same-sex marriages (Brandão & Garrido, 2023). Its demand, despite its access, has been viewed as discriminatory (Markens, 2012).

## *3.2 Legal Issues*

Since the introduction of ART in Kenya as early as 2004 (Oburu, 2016), the demand for ART to treat infertility in couples has been on the rise. Despite the high demand for these services, ART in Kenya is not regulated. There is no legal or policy regulating ART so far despite the exploitation of vulnerable couples. The ART services, embryo donation/adoption, or surrogacy in Kenya operates under the Law of Contract Act Arrangement. These contracts are managed by the clinics offering these services. Since legal conflicts will always arise from any of the ART services used, an arrangement and a contract are made and signed between the participants (intended parents, surrogates, and embryo donor) (Brandão & Garrido, 2023). The contract has several points that need to be agreed to avoid future litigation (Hageu bureau, 2012). Some of these issues include legal fees, agreements on prenatal and delivery issues, compensation, fees, insurance, and assumptions of risks that could trigger serious court cases (Crockin et al., 2020). With the lack of legal regulations, the infertility facilities in the country rely on their internal protocols and guidelines. In these guidelines, the intended parents, surrogate, or embryo donor was informed, and their rights and obligations explained. Any legal disputes that could arise are confirmed by signing the consent form.

However, legally recognized parents remain an issue and Kenyans rely on the Children Act, Cap 141, which states that one can only become a child's parent either by birth or adoption. Based on this act, the surrogate mother will remain the mother of the surrogate child until the intended parents undergo the adoptive process. As a significant step toward recognition and legalization of surrogacy in Kenya, in the year 2014, the In-Vitro Fertilization Bill was tabled in the National Assembly and later renamed as Assisted Reproductive Technology Bill; however, it was not passed (Kenya Gazette, 2014). In the year 2022, the bill was reintroduced with children born from ART considerations. The bill proposed the introduction of a surrogacy contract and the capacity of the parties, consent, rights, and obligations, and signatures of the parties involved. The flipside is that offspring from ARTs arrangements are the most important party, yet they are not in a position to sign the contract or agreement. All over across the globe, children are accepted regardless of their conception as long as their national and international human rights are upheld (Assembly, 1948). Though more than one party could be involved in motherhood, majority of countries recognized legal motherhood by birth. Therefore, giving birth is an automatic proof but for fatherhood, this could depend on the circumstances and the couple's marital status. This means a man can voluntarily acknowledge legal paternity and the birth certificate given. However, in other countries, genetic proof might undo this process. Another challenge is when a newborn child is registered in the name and motherhood rights of the surrogate mother. This could occur in Kenya, where there is no legislation governing it, allowing authorities to amend the names. But in countries where surrogacy is regulated by law and the surrogate mother's rights waived, this would be illegal and not allowed to occur (Johnson, 2017). It is observed that the ART surrogacy process could trigger litigation issues arising from any point within the process; either at pregnancy complications stage leading to its interruption, divorce or separation from intended parents, or change of mind that could trigger litigation issues (Urquiza et al., 2014). Surrogacy in Kenya is unregulated even in prior contracts, and there is no explicit legislation to follow even when the surrogate mother declines to assert her motherhood rights.

## 4 Conclusion

From the ethical issues raised from the Kenyan perspective, access to ART remains a major challenge due to poverty. The regulations of ART, safety, exploitation of the infertile couples seeking ART services, and rights of the involved parties are additional ethical dilemmas that need to be addressed. It is therefore critical to enact the pending 2022 reproductive health bill into law to regulate ART and a policy to address ART access to all (Table 9.1).

**Table 9.1** Summary of relevant papers on ethics and legal issues of surrogacy and donated embryos in Kenya

| Authors/year of publication | Research design | Methodology used | Ethical themes identified |
|---|---|---|---|
| Joan Oburu (2016) | Descriptive | Thesis review | Legal issues with rights to parenthood |
| Rashid (2018) | Descriptive | Thesis review | Need for national regulation guidelines to guide ART<br>Legal issues with rights to parenthood |
| Parkar et al. (2017) | Descriptive | Review | Informed consent |
| Kindo et al. (2020) | Descriptive | Review | Human embryo manipulation, physical and psychological trauma<br>Possible women exploitation<br>Commercial surrogacy: use of children as mere market and commercial products<br>Forced emotional detachment of mother and child after birth<br>Challenge to the structure of traditional nuclear family<br>Children born out of surrogacy facing an identity crisis<br>Lack of legislation or national regulation guidelines |

**Financial Support and Sponsorship** Nil.

**Conflicts of Interest** There are no conflicts of interest.

# References

Adamson, D. (2005). Regulation of assisted reproductive technologies in the United States. *Family Law Quarterly, 39*, 727.

Agnafors, M. (2014). The harm argument against surrogacy revisited: Two versions not to forget. *Medicine, Health Care, and Philosophy, 17*(03), 357–363.

Alizadeh, L., & Samani, R. O. (2014). Using fertile couples as embryo donors: An ethical dilemma. *Iranian Journal of Reproductive Medicine, 12*(3), 169.

Assembly, U. G. (1948). Universal declaration of human rights. *UN General Assembly, 302*(2), 14–25.

Brandão, P., & Garrido, N. (2023). Commercial surrogacy: An overview. *Revista Brasileira de Ginecologia e Obstetrícia, 44*, 1141–1158.

Brezina, P. R., & Zhao, Y. (2012). The ethical, legal, and social issues impacted by modern assisted reproductive technologies. *Obstetrics and Gynecology International, 2012*(1), 686253.

Brinsden, P. R. (2003). Gestational surrogacy. *Human Reproduction Update, 9*(05), 483–491.

Chambers, G. M., Sullivan, E. A., Ishihara, O., Chapman, M. G., & Adamson, G. D. (2009). The economic impact of assisted reproductive technology: A review of selected developed countries. *Fertility and Sterility, 91*(6), 2281–2294.

Childress, J. F., & Beauchamp, T. L. (1994). *Principles of biomedical ethics* (pp. 197–199). Oxford University Press.

Crockin, S. L., Edmonds, M. A., & Altman, A. (2020). Legal principles and essential surrogacy cases every practitioner should know. *Fertility and Sterility, 113*(5), 908–915.

Ekberg, M. E. (2014). Ethical, legal and social issues to consider when designing a surrogacy law. *Journal of Law and Medicine, 21*(03), 728–738.

Frith, L., & Blyth, E. (2014). Assisted reproductive technology in the USA: Is more regulation needed? *Reproductive Biomedicine Online, 29*(4), 516–523.

Gore, A. C., Chappell, V. A., Fenton, S. E., et al. (2015). EDC-2: The endocrine society's second scientific statement on endocrine-disrupting chemicals. *Endocrine Reviews, 36*(6), E1–E150. https://doi.org/10.1210/er.2015-1010

Gracias, C. S. (2021). Reproductive surrogacy in Ireland – An ethical and legal context. *Irish Journal of Medical Science, 190*(03), 1063–1070.

Gunnarsson Payne, J., Korolczuk, E., & Mezinska, S. (2020). Surrogacy relationships: A critical interpretative review. *Upsala Journal of Medical Sciences, 125*(2), 183–191.

Hague Conference on Private International Law. Permanent Bureau. (2012). *A preliminary report on the issues arising from international surrogacy arrangements*. Permanente Bureau.

Hodson, N., Townley, L., & Earp, B. D. (2019). Removing harmful options: The law and ethics of international commercial surrogacy. *Medical Law Review, 27*(04), 597–622.

Jadva, V. (2020). Postdelivery adjustment of gestational carriers, intended parents, and their children. *Fertility and Sterility, 113*(05), 903–907.

Jadva, V., Murray, C., Lycett, E., MacCallum, F., & Golombok, S. (2003). Surrogacy: The experiences of surrogate mothers. *Human Reproduction, 18*(10), 2196–2204.

Johnson, K. M. (2017). Contingent maternities? Maternal claims-making in third party reproduction. *Sociology of Health & Illness, 39*(08), 1349–1364.

Kenyan Gazette Notice. (2014). *The Reproductive Health Care Bill*. (2014). Kenya Gazette Supplement, Special Issue 57(Senate bill).

Kindo, N., Nyabul, P., & Njuguna, S. (2020). Ethical issues on IVF and surrogacy. *International Journal Science And Applied Research, 7*(7), 01–13.

Mackenzie, S. C., Wickins-Drazilova, D., & Wickins, J. (2020). The ethics of fertility treatment for same-sex male couples: Considerations for a modern fertility clinic. *European Journal of Obstetrics, Gynecology, and Reproductive Biology, 244*, 71–75.

Markens, S. (2012). The global reproductive health market: US media framings and public discourses about transnational surrogacy. *Social Science & Medicine, 74*(11), 1745–1753.

Mitra, P. (2018). Invisible women in reproductive technologies: Critical reflections. *Indian Journal of Medical Ethics, 3*(02), 113–119.

Montanari Vergallo, G. (2019). A child of two mothers: What about the father? Italian overview. *Acta Biomed, 90*(03), 319–325.

Montanari Vergallo, G., Marinelli, E., di Luca, N. M., & Zaami, S. (2018). Gamete donation: Are children entitled to know their genetic origins? A comparison of opposing views. The Italian state of affairs. *European Journal of Health Law, 25*(03), 322–337.

Murage, A., Muteshi, M. C., & Githae, F. (2011). Assisted reproduction services provision in a developing country: Time to act? *Fertility and Sterility, 96*(4), 966–968.

Namavar Jahromi, B., Parsanezhad, M. E., Shomali, Z., Bakhshai, P., Alborzi, M., Moin Vaziri, N., & Anvar, Z. (2018). Ovarian hyperstimulation syndrome: A narrative review of its pathophysiology, risk factors, prevention, classification, and management. *Iranian Journal of Medical Sciences, 43*(3), 248–260.

Njogu, A., Njogu, J., Mutisya, A., et al. (2022). Experiences of infertile women pursuing treatment in Kenya: A qualitative study. *BMC Women's Health, 22*, 364.

Oburu, J. (2016). *Legal parentage in surrogacy arrangements: An immediate concern in Kenya*. Doctoral dissertation. University of Nairobi.

Ombelet, W. (2011). Global access to infertility care in developing countries: A case of human rights, equity and social justice. *Facts, Views & Vision in ObGyn, 3*(4), 257–266.

Parkar, R. B., Nguyo-Kariuki, M., & Gichuhi-Wanyoike, J. (2017). Regulation of informed consent: A focus on fertility treatments in Kenya. *East African Medical Journal, 94*(6), 420–426.
Patterson, C. (2015). Embryo adoption: Some further considerations. *The Linacre Quarterly, 82*(1), 34–48.
Pennings, G. (2011). Evaluating the welfare of the child in same-sex families. *Human Reproduction, 26*(07), 1609–1615.
Präg, P., & Mills, M. C. (2017). Assisted reproductive technology in Europe: Usage and regulation in the context of cross-border reproductive care. In *Childlessness in Europe: Contexts, causes, and consequences* (pp. 289–309). Springer.
Ragoné, H. (1996). Chasing the blood tie: Surrogate mothers, adoptive mothers and fathers. *American Ethnologist, 23*, 352–365.
Rashid, F. (2018). *The legal and regulatory framework of surrogacy in Kenya: Theory and practice* (Doctoral dissertation, University of Nairobi).
Riddle, M. P. (2022). The psychological impact of surrogacy on the families of gestational surrogates: Implications for clinical practice. *Journal of Psychosomatic Obstetrics and Gynaecology, 43*(02), 122–127.
Roupa, Z., Polikandrioti, M., Sotiropoulou, P., Faros, E., Koulouri, A., Wozniak, G., et al. (2009). Causes of infertility in women at reproductive age. *Health Science Journal, 3*(2), 80–87.
Rozée, V., Unisa, S., & de La Rochebrochard, E. (2020). The social paradoxes of commercial surrogacy in developing countries: India before the new law of 2018. *BMC Womens Health, 20*(01), 234.
Saxena, P., Mishra, A., & Malik, S. (2012). Surrogacy: Ethical and legal issues. *Indian Journal of Community Medicine: Official Publication of Indian Association of Preventive & Social Medicine, 37*(4), 211.
Segal, T. R., & Giudice, L. C. (2019). Before the beginning: Environmental exposures and reproductive and obstetrical outcomes. *Fertility and Sterility, 112*(4), 613–621.
Sharma, R. S. (2014). Social, ethical, medical & legal aspects of surrogacy: An Indian scenario. *The Indian Journal of Medical Research, 140*(Suppl), S13–S16.
Shreffler, K. M., Johnson, D. R., & Scheuble, L. K. (2010). Ethical problems with infertility treatments: Attitudes and explanations. *The Social Science Journal, 47*(4), 731–746.
Sifris, R. (2015). Commercial surrogacy and the human right to autonomy. *Journal of Law and Medicine, 23*(02), 365–377.
Smith, M. K., Willmott, L., Trowse, P., & White, B. (2013). Back to the future: Prohibiting surrogacy for singles, same-sex and shorter-term heterosexual couples in Queensland. *Journal of Law and Medicine, 20*(03), 638–654.
Tabong, P. T. N., & Adongo, P. B. (2013). Infertility and childlessness: A qualitative study of the experiences of infertile couples in northern Ghana. *BMC Pregnancy and Childbirth, 13*(1), 1–10.
Urquiza, M. F., Carretero, I., Quaini, F. M., Inciarte, F., Pasqualini, R. A., & Pasqualini, R. S. (2014). Subrogación uterina: Aspectos médicos y jurídicos del primer caso con sustento legal en la Argentina. *Medicina (Buenos Aires), 74*(3), 233–238.
Varkey, B. (2021). Principles of clinical ethics and their application to practice. *Medical Principles and Practice, 30*(1), 17–28.
Wade, K. (2017). The regulation of surrogacy: A children's rights perspective. *Child and Family Law Quarterly, 29*(02), 113–131.
Wade, K. (2020). Reconceptualising the interest in knowing one's origins: A case for mandatory disclosure. *Medical Law Review, 28*(04), 731–752.
Wallach, E. E., & Robertson, J. A. (1995). Ethical and legal issues in human embryo donation. *Fertility and Sterility, 64*(5), 885–894.
World Health Organization (WHO). (2020). [Internet]. Available from: https://www.who.int/news-room/fact-sheets/detail/infertility
Yee, S., Hemalal, S., & Librach, C. L. (2020). "Not my child to give away": A qualitative analysis of gestational surrogates' experiences. *Women and Birth, 33*(03), e256–e265.

Zegers-Hochschild, F., Adamson, G. D., de Mouzon, J., Ishihara, O., Mansour, R., Nygren, K., et al. (2009). International Committee for Monitoring Assisted Reproductive Technology (ICMART) and the World Health Organization (WHO) revised glossary of ART terminology, 2009. *Fertility and Sterility, 92*(5), 1520–1524.

Zegers-Hochschild, F., Dickens, B. M., & Dughman-Manzur, S. (2013). Human rights to in vitro fertilization. *International Journal of Gynecology & Obstetrics, 123*(1), 86–89.

# Chapter 10
# Possible Violations, Exploitation, and Abuse of Surrogates in Kenya: Need for a Legislative Framework

**Teresia Nyawira Kaniaru**

**Abstract** A couple has a right to self-determination, and available technologies can be used to assist such couples who may be unable to conceive for various reasons. The principle of justice in the promotion of health ought to be considered to help the ill or disabled address their unique needs. For this to happen, it is important to have a policy and regulatory environment that is supportive of deploying relevant technologies. Assisted reproductive technologies (ART) have been used to remedy the problem of an infertile couple wishing to have a child. Depending on the cause of infertility, embryo adoption and surrogate motherhood are some of the considerations couples can make in their efforts to have a child. The absence of a legal instrument to regulate ART and surrogacy in Kenya exposes this practice to the possibility of unethical practices. Yet, the Constitution of Kenya can be seen as a tool that champions surrogacy by making it a right to form a family, a right to reproductive health, and the best interest of a child. The legislature is expected to unpack such constitutional requirements by enacting laws for purposes of regulating reproductive rights. Failure by Kenya to come up with such laws that confront the realities and complexities of sexual reproduction has exposed her vulnerable population to exploitation and abuse. This is exacerbated by systemic corruption, poverty, and other forms of deprivation among a majority of Kenyans, making surrogacy for poor women an available option for income generation and exposing them to reproductive exploitation. The objective of this chapter is to highlight some of the possible violations, exploitations, and abuse that Kenyans may experience due to the lack of a legislative tool to regulate surrogacy in the country. Some of the highlights considered include economic servitude, violations related to human dignity, the right to health, exploitation, and the possibility of supply of women and children to trafficking syndicates. The Kenyan courts have continually urged the legislature to develop a legal framework on surrogacy to avoid conflicts and exploitative tendencies which are currently common and to ensure children born out of surrogacy arrangements have their rights protected and respected.

---

T. N. Kaniaru (✉)
National Commission for Science Technology and Innovation, Nairobi, Kenya
e-mail: teresia.nyawira@nacosti.go.ke

 
L. Omutoko, W. Jaoko (eds.), *Bioethics from the Global South*, Advancing Global Bioethics 21, https://doi.org/10.1007/978-3-031-77669-4_10

**Keywords** Kenya · Legislation · Surrogacy · Reproductive health · Assisted reproductive technologies

## 1 Introduction

There is a general assumption that unprotected sexual intercourse leads to fertilization, conception, and birth. However, this is not always the case, as is in the case of infertility. Infertility has been defined as "inability of a sexually active non-contracepting couple to achieve pregnancy within one year" (Zegers-Hochschild et al., 2009). There are various causes or combinations of factors that lead to infertility, which may include ovulatory dysfunction, damaged fallopian tubes, uterine dysfunction leading to unsuccessful implantation or inability to carry a pregnancy to term, poor quality sperms, challenges related to copulation such as erectile dysfunction in men, and even genetic defects causing hormonal imbalances (Blundell, 2007). Infertility may affect one or both partners.

In an African setting, procreation is valued at both personal and community level. Moreover, the value of children extends to the personal well-being of the parents, and involuntary childlessness has negative social repercussions. Infertility has been reported as a main cause for marital instability, alcohol abuse, and sexual dissatisfaction among others (Silke, 2007).

Children are seen as offering social security, conferring social status, and satisfying their parents emotional needs (Silke, 2007). The need to have children in an African family setting cannot be overemphasized. Medical technologies have, however, provided a way out for infertile couples.

Assisted reproductive technologies (ART) have been used to remedy the problem of infertile couples wishing to have a child. Depending on the cause of infertility, embryo adoption and surrogate motherhood are some of the considerations couples can make in their efforts to have a child. Surrogacy has been defined as an arrangement where a woman agrees and bears a child for another person(s) and hands over the child to the contracted party upon birth. Surrogacy may be done for altruistic or commercial reasons. Commercial surrogacy has normalized the notion of "wombs for hire or baby factories" that have led to reproductive exploitation of the poor, making reproductive justice a mirage (Brandão & Garrido, 2022). For this reason, many countries have legal frameworks guiding ART and surrogacy. For example, commercial surrogacy is legal in India and Ukraine, while it is illegal in England, Australia, Thailand, and many states in the United States where only altruistic surrogacy is recognized (Blundell, 2007). In other countries, such as Germany and Italy, no form of surrogacy is recognized (Blundell, 2007).

As in many other countries, ART are being used in Kenya to have children. In 2006, the first child in Kenya was born out of use of ART, and in 2007, the first child was born out of surrogacy (Owuor, 2018). Interesting to note is that currently, Kenya does not have a legislative or policy framework to guide the use of ART (Olobo-Lalobo, 2022). Though there is no explicit ban, there is no formal recognition of embryo adoption or surrogacy. This makes Kenya a suitable destination for commercial surrogacy, exposing the practice to corruption, backstreet practices, self-regulation that is tainted with issues of self-interest, violations of human dignity, and other unethical practices. Failure by Kenya to confront the realities and complexities of sexual reproduction has exposed her vulnerable population to exploitation (Olobo-Lalobo, 2022). In this chapter, I will explore some of the possible violations of moral values and principles and the risks to human health and life emanating from the lack of a legal framework to address issues of ART in Kenya.

## 2 Methodology

This is a chapter that has resulted from a Google search to gather relevant information regarding the status of ART use and surrogacy practice in Kenya. Keywords were used to access this information from publications available online.

## 3 A Situational Analysis of the Legal Framework

The Constitution of Kenya (2010) can be seen as a tool that champions for surrogacy by making it a right to form a family, the right to reproductive health, and having the best interest of a child. These constitutional positions protect both intending parents and children born out of surrogacy. The Constitution has therefore provided a firm foundation upon which legislative, institutional, and policies on ART can be anchored. Since 2014, legislative attempts have been made to enact reproductive healthcare laws in Kenya but have not been successful. The Reproductive Health Care Bill (2014), In Vitro Fertilization Bill (2014), Reproductive Health Care Bill (2019), and The Assisted Reproductive Technology Bill (2019) have not seen the right of day in Kenyan parliament. Other legislative tools that would be critical in regulation of important aspects of ART have not been helpful in ensuring that ART is practiced in an ethical manner. The Children Act Cap 141 Laws of Kenya (2021), for example, is not specific to children born out of surrogacy arrangements and does not provide for legal parentage for commissioning parents (Petition 443 of 2014). In fact, it seems the Act does not even contemplate that a child can be born through surrogates. According to this Act, legal parentage can be acquired through marriage, adoption, or entering into an agreement for parental responsibility.

The other important tool is the Birth and Death Registration Act Cap 149 Laws of Kenya (1972). This Act defines a legal mother as the one who carries a pregnancy and bears a child. A notice of birth is done under the name of the mother, place, date of birth, and sex of the child. The name of the surrogate mother is captured as the legal mother, since she is the one who carries the pregnancy and gives birth (Petition

443 of 2014). Understandably, this Act was enacted in 1928 and lastly amended in 1990. Much of the dynamics facing ART were not there at that time. Commissioning parents cannot acquire a birth certificate for their child and would have to follow the process of adoption to acquire their own child who is born through surrogacy. This has become more complex after the government banned child adoption by foreigners in 2019. The Act has no place for surrogacy arrangements or ART and does not contemplate realities related to the practice or related technologies. To navigate these legislative inadequacies, hospitals have been using internal institutional mechanisms for self-regulation, while courts have been left to social engineering to rule on issues of surrogacy. The Kenyan courts have urged the legislature to develop a legal framework on surrogacy to avoid conflicts which are currently common and to ensure children born out of surrogacy arrangements have their rights protected and respected (Olobo-Lalobo, 2022).

## 4 Possible Violation, Exploitation, and Abuse of the Kenyan Population Engaged in Surrogacy Practice

It is well known that poverty functionally limits freedoms and autonomy in the practical sense and is therefore important to regulate the international surrogacy market because of its potential for abuse and exploitation of poor populations (Knoche, 2014). Surrogacy is known to flourish in countries which are poor, experience social and economic inequalities, patriarchal societies where women are viewed as inferior, and in cultures where everything can be offered for sale (Rozée et al., 2020). The Kenya National Bureau of Statistics in their Comprehensive Poverty Report published in 2020 revealed the poverty levels in the different demographics in the country. They gave the monetary poverty line as Kshs. 3252 and Kshs. 5995 in rural and urban areas, respectively, and reported that more than 36% of Kenyans are monetary poor, with youth accounting for 22% and 40% being poverty incidences in rural areas (KNBS, 2020). Over 27% of Kenyans are not only poor in monetary but in multidimensional terms because of deprivation of critical necessities. Of particular interest are 48% of the youth aged 18–34 years (6.4 million) who are classified as multidimensionally poor, with 50% being women. The poverty levels and forms of deprivation among majority of Kenyans make commercial surrogacy for the poor women an available option for income generation, exposing them to reproductive exploitation. Brandão and Garrido (2022), in their study on an overview of commercial surrogacy, identified Kenya as a popular destination for surrogacy due to the practice being less costly and the fact that there exists a legislative vacuum that can be exploited.

Commercial surrogacy in Kenya, just like in many developing countries, is practiced in a way that largely reduces surrogates to items of commerce, an object for use to satisfy well-funded market demands (Okwemba, 2012), normally negotiated by brokers and mediation agencies. The interests of the parties involved are different, and in most cases, antagonistic (Del Savio & Cavaliere, 2016). For the commissioning party, they are interested in a healthy child and engage a surrogate in a way they consider best toward this outcome. They are interested in retaining power to decide on important issues related to the process, such as the number of embryos to be implanted, whether there will be embryo reduction, whether there will be an abortion, especially if there are reports that the child might not be healthy, and even whether birth will be natural or by caesarian section. The interests of the surrogate are different. First, a surrogate under commercial surrogacy desires maximum financial benefits. They want to have control over their lives during the gestational period as well as reduce emotional attachment to the child. They are also interested in reducing the risks involved, for example, the risks that accompany hyperstimulation and multiple embryo implantations. For the brokers, their sole aim is to reap maximum financial benefits from the arrangement and to minimize possible chances of legal confrontations.

With a background of poverty, unfulfilled basic needs, and desire for a better life, intending surrogates are a disadvantaged party when it comes to having their interests taken care of in a contractual agreement. They have been known to take up risks, including traveling to foreign countries to offer the services, having multiple embryo implants, forced abortions, negotiated amounts not being fully settled, exploitation by recruiting agencies, and commissioning parents, among other unethical practices. Rozée et al. (2020), in a study conducted in India, reported that in commercial surrogacy, "only poor women who suffer become surrogates." A country without a pronounced position on ART exposes its women of reproductive age to unethical practice by intending parties and brokers who care less about the welfare of the women than their own interests, underscoring the need for regulation to avoid the potential exploitation of surrogate mothers and to ensure psychosocial evaluation and counseling.

One would argue that the requirement for informed consent by intending surrogates would be sufficient to ensure that their interests are taken care of. Informed consent is an obligatory tool which is supposed to ensure that people are treated with dignity and that their body integrity is maintained. The fundamental criteria for fair, informed consent should adhere to four main principles. These are competence, information, comprehension, and avoidance of undue influence in the consenting process. A person should have the capacity to decide once all relevant information is given. The person consenting should demonstrate the ability to comprehend the given information and make a voluntary decision. First, all necessary information that is important in supporting decision-making should be provided. This should include the various risks of participating as a surrogate. The surrogate ought to be informed that this is not just like any other pregnancy, but that it has moral and legal implications, from cultural and religious implications of being treated as an outcast to risks inherent to health in the medical procedures, social issues of being isolated in surrogate homes for a period, to bereavement after giving out the baby to intending parents.

Engagement as a surrogate starts by signing a surrogate contract, which in most cases is agreed a priori before even meeting the medical professionals who are key to the success of the pregnancy. A surrogacy contract normally has two main parts: waiver of parental rights and consent to go through a myriad of medical procedures, which include hormones, fertility drugs, fertilized embryo implantation, close medical supervision during pregnancy, and finally the process of giving birth. Various reports indicate acceptable surrogacy as one which is altruistic. Organ donation is given as an example of free of charge and motivated by altruistic reasons. Donors' altruism is looked at as a condition that ensures the donor does not compromise on their life and liberty. It follows that a surrogate who offers surrogacy, sacrifices their time, health, and is remunerated, does not do it for altruistic reasons but as an income-generating activity (Okwemba, 2012). They are considered to be vulnerable, and their consent is given under conditions of vulnerability. By eliminating the remuneration factor, the woman may not make the same decision. Though the intending surrogates consent to the process, their decisions are not free and without influence. Desire to improve their economic status coerces them to consent to participate as a surrogate.

Commercial surrogacy is the main type of surrogacy in Kenya, and money has been identified as a principal focus for surrogates engaging in the practice (Vertommen & Barbagallo, 2022). The interests of surrogate recruiting agencies who are mainly involved in consenting to the intending surrogates make them conceal certain risks, which unfold later during the period surrogates are actively engaged. Unfortunately, withdrawal from a surrogate agreement may have heavier implications for the surrogate than going through to completion. There may be no interest in having the potential surrogates comprehend the information given, and their decision-making may quite often be blurred by the fact that they may never otherwise get the kind of resources they get by engaging in surrogacy. The desire for economic uplifting blurs their voluntariness in giving their consent to participate as surrogates. Surrogates, therefore, may unintentionally waive the essence of informed consenting, especially regarding medical procedures that they have to be involved in, due to economic pressures and also from the preferences and interests of intending parents and recruiting agencies (Del Savio & Cavaliere, 2016). The economic, educational, and power disparities between the intending parents and the potential surrogates are quite often intimidating. The interests and welfare of surrogates are subsumed by the interests of the fetus, disembodying surrogate pregnancies from the surrogate. This in essence subverts their right to human dignity, health, and body integrity, reducing the surrogate to a biological incubator and unintentional servitude.

The other issue that is important and needs consideration is the mental health of the surrogate mother. Even under surrogacy arrangements, the mother retains the memory of every child they have gestated and given birth to. It is known that pregnancy has a lasting impact on the mother's brain architecture, and this plays a part in preparing the mother for the demands of the fetus and the child after birth. This is amid an African culture where a child does not just belong to a family but also a community. The society prepares to receive a child; immediately, a woman shows signs of pregnancy. Celebrations such as baby showers have become common in Kenyan society, where friends and relatives gather to express their joy once a woman is pregnant. They take this opportunity to shower the would-be mother with gifts and good wishes during pregnancy and at birth. A surrogate practicing in such a culture is likely to suffer from emotional disturbances and psychological trauma for not meeting societal expectations of presenting the baby to society upon its birth. In a country where there are challenges of poverty and a healthcare system that lacks adequate resources to cater for healthcare needs, including mental health, surrogacy may end up contributing toward unresolved mental health issues among the surrogate mothers.

Another critical factor that contributes to unethical practices in surrogacy is corruption. Corruption is a systemic problem affecting structures and institutions of government and has become a matter of great concern in the country (Hope, 2014). The practice has persisted because people benefit from its proceeds and therefore lack the will to end corrupt practices. In fact, Kenya has been classified among the most corrupt countries globally (Hope, 2013). Even though the Government of Kenya has a policy on zero tolerance to corruption and has indeed put in place institutions such as the Ethics and Anti-Corruption Commission to spearhead the fight against corruption and has a myriad other efforts geared toward fighting corrupt practices in the country, these efforts seem to be unsuccessful in rooting out this ill practice. There are many unethical practices that can exist in a corrupt system, and here, I discuss a few of them.

One would be tempted to think that lack of a legal framework guiding surrogacy practice in the country, a ban on international adoption of children, and legislations that are not friendly to surrogacy would make the environment unattractive for the practice. However, Everingham and Whittaker (2023) reported an increase in surrogacy arrangements within the country, with large sources of foreign clients coming from the United States and Australia. This could mainly be attributed to corrupt practices that enable commissioning parents and their agents to maneuver through the system. A case in point is where a commissioning parent is still able to obtain child registration and birth certificate in their name, either through having surrogate use the identity of the commissioning mother or corrupting government officers to change and capture detail of commissioning parents in the birth notification, even though the Birth and Death Registration Act spells out clearly that a notice of birth should capture the name of the mother (surrogate mother in this case), place of birth, sex of the child, and date of birth, details of which are also captured in the birth certificate of the child.

Another area of concern is the possibility of homosexuals being able to "have children." Kenya's Constitution only recognizes the marriage of heterosexual individuals. It states that "Every adult has the right to marry a person of the opposite sex, based on the free consent of the parties." Assisted reproductive technologies have brought in a dimension where sexual intercourse is not always necessary for a child to be born. The egg or sperm does not even have to come from a married couple. They can be sourced from different people, who may be of different races and even nationalities. What this means is that the commissioning party can be a man or a woman who is a homosexual and can be a single male or female. Koll (2011) cites examples of cases of gay persons who had entered into surrogate agreements in the United States. Though there is no case of homosexuals in Kenya who have been reported to use ART as a way of having children so far, it is still a possibility since the practice currently thrives in a corrupt environment.

Human trafficking has been classified as among the most lucrative criminal activities, rivaled by activities such as firearm and drug trafficking. Among the reasons that have been reported for human trafficking are sexual exploitation, cheap labor, and the extraction of organs. The 2011 report on Trafficking in Persons by the United States Department of State classifies Kenya in tier 2 group when it comes to human trafficking for not complying with set minimum standards for eradication of this vice (United States Department of State, 2011). The National Crime Research Centre, in their report on human trafficking in Kenya (2015), indicated that child trafficking is the most prevalent form, with the majority trafficked by foreigners. The same report indicated that poverty was found to be the main contributing factor for child trafficking, with parents exchanging their children for money to reduce their suffering. The sale of children is reported to take place in hospital maternity wards, children's homes, and intercountry adoptions.

One of the negative trajectories that ART has taken is its use as a "factory of babies" and feeding into child trafficking syndicates. One common feature between surrogacy and child sale is that in both instances, the child is given to a person who wants the child. Though surrogacy is legal in some countries and the sale of children is agreeably illegal as it objectifies a human being, surrogacy can and has been used as a pathway for child sale. There are many reports of the use of surrogacy arrangements to bring forth children who feed into the human trafficking syndicates that are spread globally. In June 2012, police in Nigeria freed 132 pregnant women who had illegally been placed in confinement. They reported that they were to surrender their newly born children at a fee. The children were to be sold for 2000–6000 dollars (Abdullah, 2019). In 2014, there were reports of a Japanese man who had, through surrogacy arrangements, given birth to 14 babies in Thailand, 4 of whom had already been smuggled out of Thailand (AFP, 2014). Most countries where child trafficking results from surrogacy arrangements are those where there is no explicit legal framework regulating surrogacy or in countries where surrogacy is legal. In Kenya, there is no legal framework for surrogacy, making it a fertile ground for child trafficking resulting from surrogacy or even the direct sale of children from parents. Child trafficking, being a criminal act, is not carried out in the open. Several media reports have pointed to child trafficking in Kenya. In 2018, a Kenyan woman who is alleged to have rented her womb to two Chinese nationals was arrested for child trafficking that resulted from the surrogacy contract (Musyoki, 2018). Country Reports on Human Rights Practices (2005) indicated that Kenya is indeed a source, transit, and destination of trafficked children. Arguably, the biggest scandal of child trafficking came to the limelight in 2004. In this scandal, a Kenyan-born televangelist Gilbert Deya based in the United Kingdom is alleged to have miraculously inspired conceptions through praying for infertile and postmenopausal women (Egede, 2011). These women would then travel to Kenya but would return with a baby. Investigations further revealed through DNA tests that of the 15 children who Deya claimed to be his biological children, only six were indeed his. In 2005, Mary, the wife of Gilbert Deya, was found guilty of stealing a newborn baby from the

Kenyatta National Hospital. Gilbert Deya was extradited to Kenya in 2017 to face charges of child trafficking, while his wife was jailed for 3 years for stealing a baby (Odhiambo, 2017). From this and other reports, Kenya is a fertile ground for child trafficking, and surrogacy arrangements have been used to conceal and feed child trafficking rackets.

## 5 Conclusion

Regulating surrogacy practice would enable needy but infertile couples to be able to enjoy reproductive justice and the fulfillment of having a child. It allows women who wish to participate as surrogates to do so while ensuring their human rights and dignity are upheld. Lack of a legislative tool for ART and, by extension, surrogacy makes this practice a fertile ground for potential violations, exploitation, and abuse of surrogates in Kenya. Commercial surrogacy has been reported to thrive among the poor, as they look to fulfill their basic needs and desire for a better life. The desire for economic uplifting coerces them to participate in a practice leading to economic servitude and commoditization. They are exposed to potential exploitation by being used as biological incubators without any safeguards, relegating their interests and welfare to the interests of recruiting agencies and intending parents, violating their right to being respected, health, and human dignity. Lack of legislative tools to regulate surrogacy has exposed the practice to corruption and underhand dealings, such as the supply of children to homosexuals and single parenthood. Systemic corruption also makes it a possibility for surrogacy to feed organized child trafficking syndicates. Instituting legislative measures would curb such malpractices and make it easy for law enforcement agencies, judiciary, healthcare institutions, professionals, and other key parties to perform their duties. The Kenyan courts have continually urged the legislature to develop a legal framework on surrogacy to avoid conflicts and exploitative tendencies which are currently common and to ensure children born out of surrogacy arrangements have their rights protected and respected.

## References

Abdullah, F. M. (2019). Legal and ethical aspects beyond commercial surrogacy: Modern form of human trafficking. *Journal of Legal, Ethical and Regulatory Issues, 22*(S1), 1–7.

AFP. (2014, August 8). *Sydney Morning Herald*. https://www.smh.com.au/world/thailand-launches-human-trafficking-probe-after-nine-surrogacy-babies-found-in-apartment20140808-1026b0.html

Blundell, R. (2007). Causes of infertility. *International Journal of Molecular Medicine and Advance Sciences, 3*(1), 63–65.

Brandão, P., & Garrido, N. (2022). Commercial surrogacy: An overview. *Revista Brasileira De Ginecologia E Obstetricia, 44*(12), 1141–1158. https://doi.org/10.1055/s-0042-1759774

Constitution of Kenya. (2010). http://kenyalaw.org/kl/index.php?id=398. Accessed 6 June 2021.
Del Savio, L., & Cavaliere, G. (2016). The problem with commercial surrogacy. A reflection on reproduction, markets and labour. *Biolaw Journal*. https://doi.org/10.15168/2284-4503-154
Egede, H. (2011). A re-appraisal of the Haringey London Borough Council v C (a child), E and another. *The Denning Law Journal, 23*(1), 202–226.
Everingham, S. G., & Whittaker, A. (2023, Spring). Trends in engagement in surrogacy by nationality 2018–2020: A survey of surrogacy agencies. *Global Reproductive Health, 8*(1), 64. https://doi.org/10.1097/GRH.0000000000000064
Hope, K. R. (2013). Tackling the corruption epidemic in Kenya: Toward a policy of more effective control. *The Journal of Social, Political and Economic Studies, 38*(3), 287–231.
Hope, K. R., Sr. (2014). Kenya's corruption problem: Causes and consequences. *Commonwealth & Comparative Politics, 52*(4), 493–512. https://doi.org/10.1080/14662043.2014.955981
Knoche, W. J. (2014). Health concerns and ethical considerations regarding international surrogacy. *International Journal of Gynecology & Obstetrics, 126*(2), 183–186. https://doi.org/10.1016/j.ijgo.2014.03.020. ISSN 0020-7292.
Koll, P. (2011). The use of the intent doctrine to expand the rights of intended homosexual male parents in surrogacy custody disputes. *Cardozo JL & Gender, 18*, 199.
Musyoki, M. (2018). *Mombasa woman arrested after selling baby to Chinese couple*. Available at https://www.kenyans.co.ke/.../31901-mombasa-woman-arrested-after-selling-baby-chi. Accessed 29 July 2024.
Odhiambo, T. A. (2017). *A critique of the role of the government in combatting human trafficking in Kenya*. Strathmore University.
Okwemba, A. (2012, 3 February). 'A surrogate mother speaks: I rented out my womb for Shs. 650,000', says the *Daily Nation*. Available at https://www.nation.co.ke. News. Visited 29 July 2024.
Olobo-Lalobo, J. H. (2022). Surrogacy Legislation and Kenya's ART Bill 2019: Reproductive Uhuru (freedom) a myth or a reality for infertile citizens? *African Journal of International and Comparative Law, 30*(1), 99–123. https://doi.org/10.3366/ajicl.2022.0396. ISSN 0954-8890.
Owuor, G. A. (2018). *Protection of surrogate mothers against exploitation of their reproductive organs in Kenya*. Doctoral dissertation, Strathmore University.
Petition 443 of 2014. AMN & 2 others v. Attorney General & 5 others eKLR, BKN, TMK. Available at kenyalaw.org caselaw > cases view > 105803. Visited 29 July 2024. https://kenyalaw.org/caselaw/cases/view/105803/
Republic of Kenya, Kenya National Bureau of Statistics. (2020). *Comprehensive poverty report 2020 children, youth, women, men & the elderly: From national to county level*. ISBN: 978-9966-102-21-8.
Rozée, V., Unisa, S., & de La Rochebrochard, E. (2020). The social paradoxes of commercial surrogacy in developing countries: India before the new law of 2018. *BMC Women's Health, 20*, 234. https://doi.org/10.1186/s12905-020-01087-2
Silke, J. D. (2007). The value of children in African countries—Insights from studies on infertility. *Journal of Psychosomatic Obstetrics & Gynecology, 28*(2), 69–77. https://doi.org/10.1080/01674820701409959
The Birth and Death Registration Act Cap 149 Laws of Kenya. (1972). http://kenyalaw.org/caselaw/cases/view/99217/. Accessed 6 June 2021.
The Children Act Cap 141 Laws of Kenya. Revised Edition. (2012). http://kenyalaw.org/kl/fileadmin/pdfdownloads/Acts/ChildrenAct_No8of2001.pdf. Accessed 6th June 2021.
United States Department of State. (2011). *Trafficking in persons report – Kenya, 27 June 2011*. Available at: https://www.refworld.org/docid/4e12ee6ec.html. Accessed 12 June 2021.
Vertommen, S., & Barbagallo, C. (2022). The in/visible wombs of the market: The dialectics of waged and unwaged reproductive labour in the global surrogacy industry. *Review of International Political Economy, 29*(6), 1945–1966. https://doi.org/10.1080/09692290.2020.1866642
Zegers-Hochschild, F., Adamson, F., De Mouzon, G. D., Ishihara, J., Mansour, O., Nygren, R., et al. (2009). International Committee for Monitoring Assisted Reproductive Technologies (ICMART) and the World Health Organization (WHO) revised glossary of ART terminology 2009. *Fertility and Sterility, 92*(5), 1529–1524.

# Part IV
# Global Justice

# Chapter 11
# Medical Migration in Africa: Stemming the Tide of Brain Drain

**Karori Mbũgua**

**Abstract** This chapter focuses on the migration of skilled healthcare workers from Africa to the rich countries of the Global North, a phenomenon known as the medical brain drain. I examine this phenomenon from an African moral perspective, a viewpoint that is often neglected in bioethical debates. I argue that medical migration is a betrayal of the African notions of personhood, community, and individual responsibility. However, I also maintain that appealing to African moral values alone is not likely to stop the migration of healthcare workers from the continent. Medical migration is demand driven, and the focus must shift away from the individual migrants to the destination countries themselves, as the latter bear more moral responsibility for exacerbating the already fragile healthcare systems in African countries. To promote global justice in health, the recruitment of doctors from Africa by rich nations must be criminalized in the destination countries themselves. The latter have a responsibility to avoid harming the vulnerable in Africa.

**Keywords** African ethics · Healthcare workers · Human rights · Medical migration · Vulnerability

## 1 Introduction

The history of mankind has been a history of constant migration. According to the so-called Out of Africa hypothesis, all humans evolved in East Africa and started migrating to other parts of the world over 70,000 years ago (Wilshaw, 2018). People migrate for various reasons. The reasons may be economic, social, political, or even environmental. Medical migration involves the movement of skilled healthcare workers from one geographical location to another. This may happen within a country's borders (e.g., rural to urban migration) or from one country to another. The

---

K. Mbũgua (✉)
Department of Philosophy and Religious Studies, University of Nairobi, Nairobi, Kenya
e-mail: karori.mbugua@uonbi.ac.ke

 
L. Omutoko, W. Jaoko (eds.), *Bioethics from the Global South*, Advancing Global Bioethics 21, https://doi.org/10.1007/978-3-031-77669-4_11

migration of workers affects virtually all sectors of society, but it can be particularly damaging when those migrating are medical professionals. Although Africa has the highest burden of disease in the world, many of her skilled healthcare workers educated using taxpayers' money have continued to migrate to wealthier countries such as the United Kingdom, Australia, and Canada. This migration has been putting strain on the already fragile healthcare systems of African countries, resulting in undiagnosed diseases, untreated patients, and premature deaths.

According to the World Health Organization (WHO), Africa has only 1% of the world's health workers, yet it bears 24% of the world's burden of disease (WHO, 2006). It is also estimated that 22 out of every 1000 healthcare workers in the UK National Health Service (NHS) originated from Africa (Baker, 2020). A study carried out in 2017 revealed that the number of African-trained doctors migrating to the United States had risen from 10,684 in 2005 to 13,584 in 2015 (Duvivier et al., 2017). A recent investigative report by the BBC revealed that a British company was recruiting Nigerian doctors to work in hospitals in the United Kingdom under exploitative terms not approved by the NHS (Kenyon & Meisel, 2022).

In this chapter, I will argue that the migration of skilled African medical professionals is a violation of the African notions of personhood, community, and individual responsibility. Of course, this line of thinking is not entirely novel. It was alluded to by Metz (2017) in his analysis of the debate between Gillian Brock and Michael Blake (Brock & Blake, 2015) concerning medical brain drain. In that paper, Metz argued that from an African moral perspective, it would be wrong for newly trained medical professionals to emigrate, and the state can justifiably impose emigration restrictions on such professionals. However, the current chapter goes further than this and argues that appealing to African ethics alone will not help stem the tide of medical brain drain from the continent, as Western individualistic culture has tremendously impacted the African traditional society and its way of life. The focus must shift away from the individual emigrants to the destination countries, and recruitment of highly skilled healthcare workers from Africa must be criminalized. Destination countries have a negative duty not to harm vulnerable populations, many of whom reside in Africa.

## 2 Why Medical Migration from Africa Is Morally Wrong

The migration of healthcare workers, especially doctors, from Africa to wealthier countries is a cause for moral concern because Africa has the highest burden of disease and the weakest healthcare system in the world. This migration not only causes poor health but also contributes to premature deaths in Africa. As already noted, the health worker and patient ratio in Africa is heavily tilted toward patients, and because of this, the migration of even a few health personnel can only aggravate an already bad situation. A recent survey of 47 African countries reported that the region has only 1.55 health workers per 1000 people, against the WHO recommended ratio of 4.45:1000 (Ahmat et al., 2022). The problem is compounded by the fact that Africa, with a population of nearly 1.5 billion, has only 444 medical schools, some of which were only established recently (Rosenberg et al., 2024). In 2001, it was reported that 50% of faculty members at the University of Ghana Medical School had left the country, leaving behind a vacuum (El-Khawas, 2004). In 2006, the cost of training a doctor up to university was about US$ 65,997, while that of training a nurse was 43,180 (Kirigia et al., 2006). Furthermore, it takes up to 6 years to train a doctor at the university. This means that poor African countries such as Kenya have been training health personnel to support the health services of developed countries. Indeed, this is what Bomba (2009) described as "reverse foreign aid," a situation where the poor countries are subsidizing the rich nations of the Global North. The net result of this unethical practice is an exacerbation of global health inequalities.

It has been suggested that the remittances sent back home by migrant workers may offset the negative impact of migration. Whereas it is true that migrant health workers, like all migrant workers, do send part of their savings back home, these remittances are too small to offset the high cost of training healthcare workers, especially doctors. In any case, the remittances are often sent to family members and are not ploughed back into the healthcare system of the emigrant's home country (Bradby, 2013; Chikanda & Dodson, 2013; Attutis et al., 2014). The only two developing countries that could be benefiting from their healthcare workers' remittances are Cuba and the Philippines, which have an excess supply of doctors and nurses, respectively, and which train this cadre of healthcare workers for the global market (Hooper, 2008; Brush, 2008, 2010).

## 3 African Moral Philosophy and Medical Migration

In the traditional African context, individual welfare is intimately connected to the community, and emphasis is placed on the collective good rather than the individual good (Mbiti, 1969). Accordingly, the individual does not exist outside the group and only becomes a person through social incorporation characterized by rituals and ceremonies. Moreover, as Menkiti notes, moral personhood is attained in proportion as one participates in communal life and as one discharges their obligations to the community (Menkiti, 1984: 176). Masolo (2010) makes the same point but emphasizes that the community and the individual are interdependent. In his own words, "…personhood is constituted by the various roles that people play in making the community real; individuals and community regulate and depend on each other for who and what they become" (Masolo, 2010: 218).

This means that as a person grows older, so does their personhood increase depending on whether they fulfill his role in society. Indeed, among the Gikuyu of Kenya, it is not uncommon to hear such utterances as "Ucio ti mundu" (that is not a person) when referring to an individual whose character falls short of their community's expectations. The reason for this is that in the traditional African culture, unlike in the Western culture, personhood is relational and not subjectivist. It is by relating to others that one realizes their full personhood. And as Ogbujah observes:

> The mark of a true African is his ability to put communal interests over and above his individual interests. Even when the community sponsors her talented sons in education or business, she sponsors them so that they will use their enhanced social status for the benefit of the community. When a member of the family or clan is honored or successful, the whole group rejoices and shares in the glory, not only psychologically, as one would rejoice when one's local team has won a match, but ontologically: each member of the group is really part of the honor. (Ogbujah, 2007: 14)

The Swahili proverb, "Mtu ni watu," which translates to "a person is people" in English, aptly captures the relationship between the individual and community in African thought. Given that medical education is financed by taxpayers and considering that Africa has a critical shortage of health professionals, African medical doctors have an even greater moral responsibility than the other professionals toward their country of training, and they ought not to migrate.

It is also worth mentioning that in the traditional African culture, good health is not an individual affair but a communal concern. When an individual is sick, the whole community is sick. As Omonzejele aptly puts it:

> For the traditional African, health is not just about the proper functioning of bodily organs. Good health for the African consists of mental, physical, spiritual, and emotional stability [of] oneself, family members, and community; this integrated view of health is based on the African unitary view of reality. Good health for an African is not a subjective affair. (Omonzejele, 2008: 120)

Indeed, in traditional Africa, caring for the sick was never considered a favor but a moral duty (Chuwa, 2014). It was every individual's duty to take care of sick family members, relatives, and members of the larger community, and those who failed to do so were admonished (Osuji, 2013). In traditional Africa, the immediate members of the community went out of their way to mobilize resources to help the sick and those in need (Tangwa, 2005). This practice continues up to the present day in the form of fundraisers organized to offset medical bills for relatives, friends, neighbors, workmates, and even strangers detained in hospitals over unpaid bills.

Furthermore, traditional healers were paid very little for services rendered, and full payment was only made after the patient had been healed (Leonard, 2000). Because of the important role that these healers played in the community, they occupied an important place in the social hierarchy alongside elders and chiefs (Yalae, 2008). Today, however, it is worth mentioning that in some African countries, such as Kenya, doctors receive higher pay compared to other professionals (Kirigia et al., 2006). This is an indication that the services that doctors render are still very highly valued.

Because medical education in Africa is mainly financed by the public, Africa-trained doctors have a special moral obligation to give back to the community upon successful completion of training. Besides, locally trained African doctors are in a better position to give medical care to local patients as they share the same language and culture. It is also worth mentioning that even in the West, altruism or self-sacrifice has been a core component of the practice of medicine since the time of

Hippocrates (Feldman, 2017; Jones, 2002). Indeed, that is why medicine is often viewed as a noble calling rather than a mere trade or a money-making endeavor.

## 4 Medical Migration and Human Rights

Liberalism believes that society is made of autonomous individuals capable of making their own decisions. In this view, the individual rather than the community is the originator of values, and their rights must never be trampled upon. A liberal might therefore argue that curtailing medical migration is wrong because it undermines the freedom of movement enshrined in the Universal Declaration of Human Rights (UDHR). Article 13(2) states that "everyone has a right to leave any country, including his own." Carrens underscored the right of the individual to migrate when he said that "to keep people from emigrating (by denying them a place to go) because they represent a valuable resource to their country of origin would be a dramatic departure from the liberal tradition" (1987: 261).

However, Article 29 of the UDHR also explicitly states that "*everyone has duties to the community* in which alone the free and full development of his personality is possible." Sadly, this feature of the declaration is often overlooked. As already indicated, the emigration of skilled healthcare workers from developing to developed countries undermines the healthcare systems of the sending countries and contributes to global injustice in health.

Medical migrants sometimes find themselves jobless upon reaching their destination countries. Others end up taking lower jobs or jobs they never trained for, a phenomenon referred to as brain waste or deskilling. As one medical immigrant, as quoted by Lofters et al. (2014), remarked:

> ....issuing a visa to an IMG (International Medical Graduate) like me to work as a security guard is simply a mockery of the immigration dream. I immigrated to ensure a secure future for my two daughters who lost their mother in 2009. By immigrating to be a security guard at the age of 52, I have betrayed my poor nation, which made me a doctor with taxpayers' money. Also, by throwing into the garbage my long 25 years of experience and knowledge of medicine I have committed a crime to myself. [Male, aged 50+ years]. (Lofters et al., 2014: 86)

It is also worth mentioning that most migrant doctors, at least in the United Kingdom, are disproportionately concentrated in certain specialties such as psychiatry, geriatrics, and general practice, which are often considered less prestigious or dirty and avoided by locals (Simpson, 2014).

Even moderate communitarians like Kwame Gyekye recognize that rights are subject to reasonable restrictions. As he puts it:

> Individual rights, the exercise of which is meaningful only within the context of human society, *must therefore be matched with social responsibilities*. In the absence of a display of sensitivity to such responsibilities, the community will have to take the steps necessary to maintain its integrity and stability. The steps are likely to involve abridging individual rights... (Gyekye, 1997: 65)

In another place, Gyekye says:

> The social and ethical values of social well-being, solidarity, interdependence, cooperation, compassion, and, reciprocity, which can be said to characterize communitarian morality, primarily impose on the individual duty to the community and its members. It is all these considerations that elevate *the notion of duties to a priority status in the whole enterprise of communitarian life.* (Gyekye, 1992: 119)

It should be clear by now that the traditional African worldview with its emphasis on communal values and relationships is in direct opposition to Western liberal thought, which emphasizes the freedom of the individual. Liberalism is incapable of addressing the problem of brain drain in Africa precisely because it places high value on individual self-determination but fails to recognize that man is social by nature and craves to belong to a community. It also overlooks the alienation and estrangement that ensues when this desire is unfulfilled (Wellman, 1999; Sandel, 1982). It fails to take into account the fact that an individual requires the cooperation of others to survive and thrive.

Because liberalism emphasizes rights over duties, its approach to human rights is at odds with the African conception of human rights. From an African perspective, individual rights are supposed to complement collective rights. For this reason, the African approach to human rights, unlike the Western conception, puts more premium on duties over rights. Indeed, as Ikuenobe has pointed out, undue emphasis on rights over duties is not a good way of ensuring the rights of all, especially the disadvantaged. Dissatisfied with the Universal Declaration of Human Rights, which he accuses of value and cultural imperialism, he has gone on to call for a Universal Declaration of Responsibility (Ikuinobe, 2018).

In collectivist cultures and in situations where there is a serious shortage of health personnel, as is the case in most African countries, thoroughgoing liberalism is not desirable, and the decision to leave a country that invested in one's education cannot be morally justified. Indeed, a human rights approach that insists on an individual's absolute right to migrate is incompatible with African communal values. This is not to say individuals did not have rights in traditional African society. On the contrary, they did, but these rights were exercised in such a way as not to undermine the collective good. In developing countries with an inadequate number of health workers, limiting the migration of such workers by bonding recipients of government scholarships (through legislation) would be morally unobjectionable.

Unfortunately, appealing to African traditional values *alone* will not stop the migration of doctors from Africa without the corresponding cooperation of the receiving countries themselves. The reality of the matter is that Africa is not free from Western cultural influence. Many Africans have embraced Western values and become individualistic, thanks to neo-liberalism, the economic orthodoxy that now dominates the world today. As Stuart Rennie graphically pointed out, "Contemporary 'African life' is not predominantly lived in villages led by traditional elders where communal problems are resolved by palabre under a baobab tree. That image is becoming increasingly quaint against the influences of colonialism and globalization, increased urbanization, digital communication, and the subsequent fraying of traditional community structures" (Rennie, 2016).

It is hypocritical for Western governments to champion human rights while at the same time undermining the health systems of poor African countries. The poaching of highly skilled healthcare workers from Africa is not significantly different from the sixteenth- to nineteenth-century transatlantic slave trade in terms of its exploitative and devastating effects on Africa (Olumide & Upkere, 2012; Oforka, 2015; Heath, 2007). The only difference is that while the slave trade robbed Africa of its young and able-bodied people, the ongoing brain drain (as the name suggests) is robbing Africa of its best brains and undermining its citizens' right to health, and it should therefore be outlawed. It contradicts the principle of vulnerability, which emphasizes the need to avoid placing vulnerable people at risk of harm.

The demand for doctors in destination countries is the major pull factor for medical migration. This demand is associated with a relative shortage of health personnel and *internal* brain drain in destination countries themselves, a situation where most locally trained doctors prefer to work in certain geographical locations (often urban areas) and certain fields of medicine (Taylor & Esmail, 1999; Raghuram et al., 2009). Training more healthcare professionals in the destination countries could help reduce Africa's medical brain drain as this will reduce the demand for immigrant healthcare professionals.

Another strategy would be to criminalize (through legislation) the active recruitment of doctors from developing countries in the same way that the slave trade was criminalized in the nineteenth century. Indeed, just as attention has shifted away from the sex worker to the client in some European countries, so too must attention shift away from the individual emigrants and the sending countries to the destination countries themselves. The criminalization of active recruitment of doctors from Africa (and other poor nations) coupled with enhanced medical training in the destination countries themselves would go a long way in reducing the demand for doctors and, as a consequence, ensuring global justice in health.

## 5 Conclusion

Unchecked migration of African health workers has continued to undermine Africa's healthcare systems and worsen global health inequalities. We have seen that such migration is antithetical to the African notion of personhood, which prioritizes the common good over individual liberty. Indeed, the migration of highly skilled African health workers is a betrayal of the spirit of *ubuntu* or humanism, which is at the core of African moral thought. There can be no greater betrayal than this. But medical migration is demand driven. Therefore, there is a need to shift the focus of attention away from the sending countries (and the medical immigrants themselves) to the receiving countries. Receiving countries are literary reaping where they did not sow. Indeed, they are more morally culpable than the emigrants themselves. Receiving countries must solve their health workforce challenges without recruiting health workers from Africa, as this places vulnerable African citizens at risk of harm.

## References

Ahmat, A., Okoroafor, S. C., Kazanga, I., et al. (2022). The health workforce status in the WHO African region: Findings of a cross-sectional study. *BMJ Global Health, 7*, e008317. https://doi.org/10.1136/bmjgh-2021-008317

Attutis, C., Bishaw, T., & Frank, M. W. (2014). The workforce for health in a globalized context - global shortages and international migration. *Global Health Action*, 7(s2). https://doi.org/10.3402/gha.v7.23611@zgha20.2014.7.issue-s2

Baker, C. (2020). *NHS staff from overseas: Statistics*. UK Parliament House of Commons Library. Briefing Paper Number 7783. https://commonslibrary.parliament.uk/research-briefings/cbp-7783/

Bradby, H. (2013). *A review of research and policy documents on the international migration of physicians and nurses* (Max Plank Institute for the Study of Religious and Ethnic Diversity Working Paper No. 13-07).

Brush, B. L. (2008). Global nursing migration today. *Journal of Nursing Scholarship, 40*(1), 20–25. https://doi.org/10.1111/j.1547-5069.2007.00201.x

Brush, B. L. (2010). The potent lever of toil: Nursing development and exportation in the postcolonial Philippines. *American Journal of Public Health, 100*(9), 1572–1583.

Bomba, M. (2009). Exploring legal frameworks to mitigate the negative effects of international health-worker migration. *Boston University Law Review, 89*(103), 1104–1135.

Brock, G., & Blake, M. (2015). *Debating brain drain: May governments restrict emigration?* Oxford University Press.

Carrens, J. (1987). Aliens and citizens: The case for open borders. *Review of Politics, 49*, 251–273.

Chikanda, A., & Dodson, B. (2013). Bandaid transnationalism: Remittance practices of emigrant Zimbabwean medical doctors. *Migration and Development, 2*(1), 57–73. https://doi.org/10.1080/21632324.2013.773152

Chuwa, L. T. (2014). *African Indigenous bioethics in global bioethics: Interpreting Ubuntu*. Springer. https://doi.org/10.1007/978-94-017-8625-6

Duvivier, R., Burch, V., & Boulet, J. (2017). A comparison of physician emigration from Africa to The United States of America between 2005 and 2015. *Human Resources for Health, 15*(1), 41. https://doi.org/10.1186/s12960-017-0217-0

El-Khawas, M. (2004, Fall). Brain drain: Putting Africa between a rock and a hard place. Mediterranean Quarterly.: 37–56, 15.

Feldman, F. (2017). Altruism and medical practice. *Journal of General Internal Medicine, 32*(7), 719–720.

Gyekye, K. (1992). Person and community in African thought. In K. Gyekye & K. Wiredu (Eds.), *Person and community: Ghanaian philosophical studies* (pp. 101–122). Council for Philosophical Values.

Gyekye, K. (1997). *Tradition and modernity: Philosophical reflections on the African experience*. Oxford University Press.

Heath, I. (2007). Exploitation and apology. *British Medical Journal, 334*, 981–998.

Hooper, C. R. (2008). Adding insult to injury: The healthcare brain drain. *Journal of Medical Ethics, 34*(9), 684–687.

Ikuinobe, P. (2018). Human rights, personhood, dignity, and African communalism. *Journal Human Rights, 17*(5), 589–604. https://doi.org/10.1080/14754835.2018.153345

Jones, R. (2002, March 16). Declining altruism in medicine: Understanding medical altruism is important in workforce planning. *British Medical Journal, 324*, 624–625.

Kenyon, P., & Meisel, A. (2022, October 11). 'Exploited' foreign doctors worry about risk to UK patients. *BBC News*. www.bbc.com/news/uk-6314192

Kirigia, J., Gbary, A., Muthuri, L., et al. (2006). The cost of health professionals' brain drain in Kenya. *BMC Health Services Research, 6*, 89.

Leonard, K. (2000). *African traditional healers: The economics of healing*. Indigenous Knowledge (IK) Notes; No. 32. World Bank, Washington, DC. © World Bank. https://openknowledge.worldbank.org/handle/10986/10806. License: CC BY 3.0 IGO.
Lofters, A., Slater, M., Fumakia., & Thulien, N. (2014). "Brain drain" and "brain waste": Experiences of international students in Ontario. *Risk Management and Healthcare Policy, 7*, 81–89. https://doi.org/10.2147/RMHP.S60708
Masolo, D. (2010). *The self and community in a changing world*. Indiana University Press.
Mbiti, J. (1969). *African religions and philosophy*. East African Educational Publishers.
Menkiti, I. (1984). Person and community in African traditional thought. In R. Wright (Ed.), *African philosophy: An introduction* (pp. 171–181). University Press of America.
Metz, T. (2017). The ethics and politics of the brain drain: A communal alternative to liberal perspectives. *South African Journal of Philosophy, 36*(1), 101–114.
Oforka, V. C. (2015). *The bleeding continent: How Africa became impoverished and why it remains poor*. Xlibris.
Ogbujah, C. (2007). The individual in African communalism. *Melintas, 23*(1), 13–27.
Olumide, H., & Upkere, W. (2012). Brain drain and African development: Any possible gain from the drain? *African Journal of Business Management, 6*(7), 2421–2428.
Omonzejele, P. (2008). African concepts of health, disease, and treatment: An ethical inquiry. *Explore, 4*(2), 120–126.
Osuji, P. (2013). *The contribution of African traditional medicine for a model of relational autonomy in informed consent*. Doctoral dissertation, Duquesne University. Retrieved from https://dsc.duq.edu/etd/1000
Raghuram, P., Bornat, J., & Henry, L. (2009). Ethnic clustering among South Asian geriatricians in the UK: An oral history study. *Diversity in Health and Care, 6*, 287–296.
Rennie, S. (2016, March 16). *Is there an 'African bioethics'?* http://globalbioethics.blogspot.com/2016/03/is-there-african-bioethics.html
Rosenberg, J., Juvonen, M., Ng, M. Z., Arinzeh. N., Adanu, T. S. A., Ninsiima, H., Temesgen, R., Ajuwon, G., Abiodun-Asanre, O. A., Pellegrom, S., Weintraub, R., & Hailu, T. (2024). Medical schools in Africa: Seeing momentum. *BMJ Global Health, 9*(2), eo14468. https://doi.org/10.1136/bmjgh-2023-014468
Sandel, M. (1982). *Liberalism and the limits of justice*. Cambridge University Press.
Simpson, J. (2014, December 8). Diagnosing a flight from care: Medical migration and 'dirty work' in the NHS. *History and Policy*. Available from: https://www.historyandpolicy.org/policy-papers/papers/diagnosing-a-flight-from-caremedical-migration-and-dirty-work-in-the-nhs. Accessed 1 Aug 2022.
Tangwa, G. (2005). The HIV/AIDS pandemic, African traditional values and the search for a vaccine in Africa. In A. van Nierkerk, and L. M. Kopleman (Eds.), *Ethics and AIDS in Africa: the challenges of our thinking* (pp. 179–189). David Phillips Publishers.
Taylor, D. H., & Esmail, A. (1999). Retrospective analysis of census on general practitioners who qualified in South Asia. *BMJ, 318*(7179), 307–308.
Wellman, C. H. (1999). Liberalism, communitarianism and group rights. *Law and Philosophy, 18*(1), 13–40.
Wilshaw, A. (2018). The out of Africa hypothesis Trevathon W. In *The international encyclopedia of biological anthropology* (3 Vols.). Wiley-Blackwell.
World Health Organization. (2006). *The world health report: 2006: Working together for health*. World Health Organization. https://apps.who.int/iris/handle/10665/43432
Yalae, P. (2008). *Neo-Africanism: The new ideology for a new Africa*. Trafford Publishing.

# Glossary

**Anthropocentrism** A philosophical perspective that regards humans as separate from and superior to nature. It views human life as having intrinsic value while animals, plants, and mineral resources are merely treated as resources that can be exploited for the benefit of mankind and places humankind at the center of the world, prioritizing human interests over those of other species and the environment.

**Autonomy** An individual's capacity to make informed self-directed decisions about their own life and matters related to health, medical care, and research.

**Embryo adoption** Also known as embryo donation, it involves the transfer of cryopreserved embryos from one couple or individual to another to allow the recipient to carry and give birth to a child.

**Embryo donation** Offering of embryos to other couples or individuals struggling with infertility; embryos come from couples undergoing in vitro fertilization and have surplus embryos remaining after achieving their reproductive goals. The donation is done instead of discarding the embryos.

**Environmental ethics** A branch of philosophy focusing on understanding and addressing humans' ethical responsibilities toward the natural world. It considers questions related to the value of nature, the impact of humans on ecosystems, and the interconnectedness of living things.

**Legalization** The law's requirement to deal with constitutional issues arising in medicine and the life sciences. It ensures that the law and ethical rules guide actions in healthcare and scientific exploration.

**Legislation** The action of making something that was previously illegal permissible by law.

**Medical migration** The movement of healthcare professionals from one country to another to practice medicine.

**One health** A transdisciplinary approach for achieving optimal health outcomes that acknowledge the interconnectedness between humans, animals, plants, and their shared environment.



L. Omutoko, W. Jaoko (eds.), *Bioethics from the Global South*, Advancing Global Bioethics 21, https://doi.org/10.1007/978-3-031-77669-4

**Pandemic** A widespread occurrence of an infectious disease over a whole country or the world at a particular time that affects a significant proportion of the population and can occur across continents.

**Patient rights** A component of autonomy that includes the right to refuse treatment, the right to access medical information, and the right to participate in decisions about one's own health.

**Surrogacy** An arrangement in which a woman agrees to carry and give birth to a child on behalf of another person or couple. It involves medical, legal aspects, and deep emotional connections.

# Index




L. Omutoko, W. Jaoko (eds.), *Bioethics from the Global South*, Advancing Global Bioethics 21, https://doi.org/10.1007/978-3-031-77669-4

The manufacturer's authorised representative in the EU is Springer Nature Customer Service Centre GmbH, Europaplatz 3, 69115 Heidelberg, Germany. If you have any concerns regarding our products, please contact ProductSafety@springernature.com

Printed and bound by CPI Group (UK) Ltd, Croydon, CR0 4YY
30/09/2026
02225558-0001